E   t
Engineers

Electronics for Student Engineers

BRUCE NEWBY

Butterworth-Heinemann Ltd
Linacre House, Jordan Hill, Oxford OX2 8DP

A member of the Reed Elsevier plc group

OXFORD LONDON BOSTON
MUNICH NEW DELHI SINGAPORE SYDNEY
TOKYO TORONTO WELLINGTON

First published 1996

7 day loan

British Library Cataloguing in Publication Data
Newby, B. W. G.
  Electronic for Student Engineers
  I. Title
  621.381

ISBN 0 7506 2144 3

Library of Congress Cataloging-in-Publication Data
Newby, B. W. G. (Bruce W. G.)
  Electronics for student engineers/Bruce Newby.
  p.   cm.
  Includes index.
  ISBN 0 7506 2144 3
  1. Electronics. I. Title.
TK7816.N49   95–37488
621.3–dc20   CIP

Printed in Great Britain

Contents

Preface

This book is recommended reading for mechanical engineering students following Intermediate and Advanced level GNVQs or BTEC NC and HNC courses. It is also recommended for the electrical or electronic C&G craft student and anyone who requires a straightforward approach in a single textbook to a combination of both basic electrical principles and simple analogue electronics.

While electrical principles themselves do not change, this book attempts to present them in a new way specifically designed to appeal to the mechanical student. Some of the explanations in the book include hydraulic analogies which the author has found useful in the teaching of electronics to mechanical students. For example, in Chapter Four the bathing pool and the speedboat are used to help illustrate the difference between electrical resistance, reactance and impedance. Again, in Chapter Nine, the action of the transistor, as used in the common emitter amplifier, is likened to a water-tap controlling the flow of water through a pipe. As with all analogies, these are not precise; however, they do help the mechanically inclined student towards an understanding of the electrical principles involved.

Many circuit diagrams and other sketches are included. Where appropriate, these contain additional explanatory notes, which are emphasized by being enclosed in *note clouds*.

Finally, all the appropriate chapters include worked numerical examples and conclude with exercises for which answers are given.

Bruce Newby

1 Electrical units and components

1.1 The nature of electricity

1.1.1 Introduction

While most people are very familiar with the use of electricity, they are somewhat vague about explaining just what electricity is. They know that it can provide heat and light and can power their televisions and other electrical appliances. If they have ever had an electric shock, then they know that it can be dangerous if not treated with respect. To understand the precise nature of electricity, we need to be aware of the type of materials used in its manufacture and in its subsequent distribution.

1.1.2 Atoms and elements

If we took a piece of metal, say a piece of pure copper, and kept reducing it in size to the smallest possible, we would finish with an atom of copper. The size of the atom would be so small that we could not see it with the naked eye.

All the atoms of pure copper are identical. Such a material is known as an *element*. Iron, aluminium, carbon, sulphur, silicon, germanium, hydrogen and oxygen are further examples of elements. There are over a hundred different elements known to man.

While all atoms comprising a particular element are identical, the atoms of different elements are not alike. Just how atoms differ between elements becomes evident if we examine the structure of the atoms belonging to different elements. We shall do this in Section 1.1.4.

1.1.3 Electrification by friction

First let us establish the meaning of positive and negative as applied to electricity. Most readers will be familiar with the electrifying effect of combing dry hair. Drawing the comb through the hair causes crackling and sparking; afterwards, the comb attracts light pieces of paper. We can reproduce this effect in a very dry atmosphere by rubbing two glass rods with silk or by rubbing two ebonite rods with fur. Bringing the two treated glass rods together causes them to repel each other. Similarly, the two treated ebonite rods repel each other. However, a treated glass rod will attract a treated ebonite rod.

Clearly, this produces two forms of electrification. That on the glass rods is positive electricity, and that on the ebonite rods is negative electricity. The experiment shows that rods that are similarly charged electrically repel, while a rod that has received a positive electrical charge will attract one that has a negative electrical charge. In short, *like charges repel each other but unlike charges attract each other*. Figure 1.1 illustrates this point.

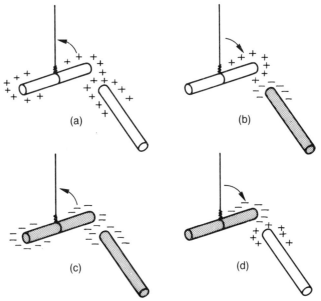

Figure 1.1 *Mutual effect of like and unlike charges. (a) Two positively charged glass rods repel; (b) a positively charged glass rod and negatively charged ebonite rod attract; (c) two negatively charged ebonite rods repel; (d) a positively charged glass rod and a negatively charged ebonite rod attract*

In the above experiment, the silk used to produce the positive electrical charge on the glass rods became negatively charged to an equal degree. Similarly, the fur used to produce the negative electrical charge on the ebonite rods became equally positively charged. This suggests that the rubbing action did not generate electricity but merely separated that which already existed. This concept fits in very nicely with the positive and negative terminology because if the two types of electricity do co-exist in equal amounts then the result will be the original electrical neutrality of the untreated rods.

1.1.4 Atomic structure

It has been known for many years that all materials are made up of vast quantities of atoms. This is true whether the material is a pure element, comprising the same type of atom throughout, or a compound or mixture of different elements. It is also a known fact that while the atoms of different elements are not the same size, they are all made up in the same way. All atoms consist of a heavy, positively charged central mass, called the nucelus, surrounded by a number of orbiting negatively charged particles called electrons. The atoms are not naturally electrically charged and their neutrality is achieved by the amount of positive electrical charge on the nucleus being balanced by the sum of the negative charges on the electrons. The structure of the atom can be likened to that of our own solar system. The sun represents the nucleus and the orbiting planets the swirling electrons. Also, just as most of the solar system is empty space, so is most of the volume of an atom!

The difference between the atoms of different materials, or elements, is not in the manner in which they are structured but in the different sizes of the nuclei and the different numbers of orbiting electrons. The individual electrons are identical in all elements; it is only their numbers that vary. In all atoms of all elements there is a natural electrical neutrality. Take, for example, an atom of hydrogen. This has a single electron orbiting about its nucleus. The helium atom has a slightly larger nucleus requiring two orbiting electrons for electrical balance. Three electrons are to be found in an atom of lithium; two are in the same orbit around the nucelus but the third is in its own orbit a little further away.

If we keep adding negative electrons, at the same time increasing the positive charge on the nucleus by the same amount, we produce a succession of different types of atoms belonging to different elements. The mechanics of the system is such that the swirling electrons tend to stay in fixed orbits. The orbit nearest to the nucleus can hold up to two electrons only; any more spill over into the second orbit which can take eight before it is full and spills over to the third orbit. The third orbit is filled by 18 electrons; the fourth by 32 and so on. The different electron orbits are called electron shells and are identified by the letters K, L, M, N, O, P and Q.

Many atoms have only sufficient electrons to occupy a few of the shells nearest to the nucleus. Hydrogen and helium use only the K-shell while lithium uses a full K-shell and only one electron in the L-shell. Silicon has a total of 14 electrons and fills both the K-shell and the L-shell leaving four electrons to spill over into its outer shell, the M-shell. Figure 1.2 shows a graphical representation of the silicon atom. The elliptical lines represent the orbit paths and the black dots the electrons in those orbits. Figure 1.3 shows another method of representing the nucleus and its surrounding electron shells for three example atoms: neon gas, silicon and copper. In this diagram the electrons are still represented by black dots but their orbits are represented by simple circles centred on the nucleus.

Now, it is a known fact that the electrons in the outer shell of any atom are not attracted so strongly to the nucleus as those in the inner shells. It is also known that if the outer shell does not have its full complement of electrons then it is possible to dislodge one such that it leaves the parent atom altogether. If, for example, the outer shell has only one electron,

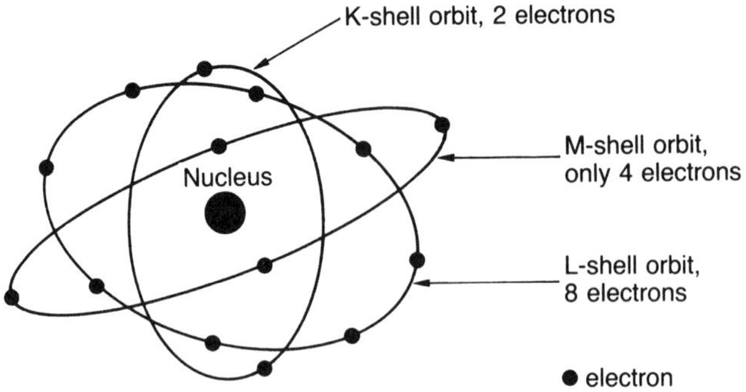

K-shell orbit, 2 electrons

Nucleus

M-shell orbit, only 4 electrons

L-shell orbit, 8 electrons

● electron

Figure 1.2 *Model of a silicon atom having a +14 positive nucleus and 14 balancing negative electrons. Only 4 electrons are in the outer M-shell which has a full shell capacity of 18 electrons.*

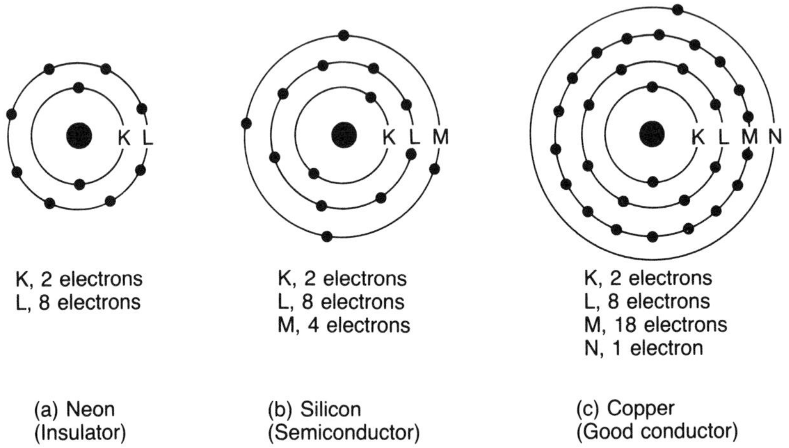

K, 2 electrons
L, 8 electrons

K, 2 electrons
L, 8 electrons
M, 4 electrons

K, 2 electrons
L, 8 electrons
M, 18 electrons
N, 1 electron

(a) Neon
(Insulator)

(b) Silicon
(Semiconductor)

(c) Copper
(Good conductor)

Figure 1.3 *Examples of graphical models for three different atom structures showing the numbers of electrons in the electron shells*

then that electron is so loosely bound to its nucleus that it is very easily dislodged to become a free electron. As such, it will wander aimlessly around the adjacent atoms which will do nothing to prevent this, since they themselves are largely free space. However, it should be remembered that if an electron leaves its parent atom then the positive nucleus is no longer balanced by its correct complement of orbiting negative electrons. This means that the atom, deficient of an electron, is effectively positively charged. This positively charged atom is said to be ionized and will have a natural tendency to attract back the free electron which it has just lost. But, if the element, of which our atom is only a tiny part, is influenced by a positive electrical charge at one end, the free negative electron will be attracted to that positive charge and will move through the element accordingly.

Figure 1.3(a) represents neon gas. This has its K-shell filled by two electrons and its outer shell, the L-shell, has eight electrons. Now the full complement of the L-shell is eight electrons so with this shell being completely filled any one of these electrons is difficult to remove. For this reason, we shall find that neon is a poor conductor of electricity; it is a good insulator.

Figure 1.3(b) is another representation of the silicon atom shown originally in Figure 1.2. Note that there are only four electrons in the M-shell, which requires 18 to fill it. At room temperature, it is very difficult to move an electron from this outer shell. At temperatures near that of boiling water, however, it is very easy indeed to cause a free electron in silicon. Silicon is called a semiconductor and is very important in electronics. We shall take a further look at silicon in Chapter 7.

Figure 1.3(c) is the model for copper. This has its inner shells K, L and M completely filled but has only one electron in its outer N-shell. As expected, this single electron is readily detachable and as we shall see below makes copper a good conductor of electricity.

1.1.5 Free electrons and electricity

From the foregoing discussion on the structure of atoms it is apparent that some materials are more likely than others to contain free electrons. The materials that have atoms with largely incomplete outer shells are the ones which will readily free an electron. Copper is one such material and others include gold, silver, aluminium, brass and most other metals. But what is it which causes an electron to break away from its parent atom and become 'free'? There are several effects which can cause a free electron but at this stage we shall concern ourselves with perhaps the most likely one. This is an applied electrical potential or, to give it its proper name, an applied electromotive force (e.m.f.). Such a source of e.m.f., measured in volts, is the simple torch battery or motor car battery. Both of these have positive and negative terminals between which the e.m.f. or voltage appears.

Now, a length of copper wire can be regarded as nothing more than a long mass of copper atoms every one of which is in physical contact with its neighbours. If a piece of copper wire is connected to a lamp and a battery, as shown by Figure 1.4, then we notice that the lamp lights. This is because the electric force of the battery (its voltage) provides the necessary extra energy to make the loosely bound outer electrons in the copper atoms break free. However, this time the freed electrons are not left to wander about at random. Being negative particles, they are immediately attracted by the positive terminal of the battery. The free electrons in the copper thus drift through the atomic structure of the copper wire to disappear into the battery positive terminal. As each electron passes from the copper wire into the battery positive terminal it is immediately replaced by an electron released from the battery negative terminal.

It is this movement of electrons through the copper material which constitutes what we call electricity. It is this same flow of electrons through the lamp of Figure 1.4 which causes it to light. Note that the direction of flow of electrons is from the battery negative terminal to the battery positive terminal.

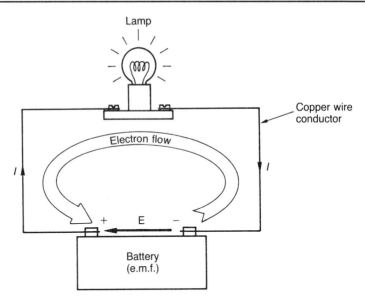

Figure 1.4 *Conventional current and electrons flow in opposite directions. Conventional current, I, flows from the battery positive terminal to the battery negative terminal. The flow of electrons is actually in the opposite direction.*

I is measured in amps and is proportional to the number of electrons per second arriving at the battery positive terminal.

E is the battery e.m.f. measured in volts

Unfortunately, long before this electron theory was properly appreciated, it had been decided by convention that electric current flowed from positive to negative. It was now too late to change the direction of flow of current to that of the electrons and so to this day we still accept that conventional current and electrons flow in opposite directions.

To summarize, we could define electricity as the flow of electrons from one particle to another caused by some outside electromotive force. The flow of electrons or electric current is measured in amperes, or amps for short.

1.2 Electrical units

1.2.1 The ampere

The ampere or amp (unit symbol, A) is a measure of electric current (I). It is approximately 6.26×10^{18} electrons per second. In other words, if in one second this number of electrons is fed into a wire conductor, that conductor would be passing a current, I, of 1 A.

1.2.2 The coulomb

The coulomb (unit symbol, C) is a measure of electrical quantity or charge (Q). Since electrons are electrical charges, the coulomb is the name given to 6.26×10^{18} electrons. So, if a current of one amp flows for one second the quantity of charge transferred is one coulomb:

Coulombs = amps × seconds

or

C = A × s

or

$$Q = I \times t \qquad\qquad [1.1]$$

1.2.3 The volt

The volt (unit symbol, V) is a measure of electric potential or voltage (V) or electric force or e.m.f. (E). The voltage difference between the two ends of a conductor is the electrical force which 'pushes' the current, amps, along the conductor. The greater the voltage applied between the ends of the conductor, the greater the current that is caused to flow along the conductor. It is as if the conductor has some natural resistance to current being made to flow through it. This is exactly the case. The conductor is said to exhibit resistance, measured in ohms (see the next section), to the flow of current.

A volt is generally taken as being the potential difference, that is, voltage, required to drive a current of one amp through a component of resistance one ohm.

1.2.4 The ohm

The ohm (unit symbol, Ω) is a measure of a component's resistance (R) to the flow of current (I) through it. For example, a conductor of small gauge wire will have a greater electrical resistance than a conductor of the same length but of a larger gauge. Clearly, the nature of the material used in the conductor and the length of the conductor will also affect its resistance. It can be shown that the electrical resistance, R, of a conductor of cross-sectional area $A\,\mathrm{m}^2$, of length 1 m and of material having a resistivity of $\rho\ \Omega$-m is:

$$R = (\rho\,I/A)\ \Omega \qquad\qquad [1.2]$$

Example 1.1 *Calculate the resistance of a copper wire conductor 3 km long and cross-sectional area 2.5 mm². Take the resistivity of copper, ρ, to be 1.6 × 10⁻⁸ Ω-m.*

$$R = 1.6 \times 10^{-8} \times 3000 \div 2.5 \div 10^{-6}$$
$$= 19.2\ \Omega$$

1.2.5 The siemen

The siemen (unit symbol, S) is a measure of electrical conductance (G). The conductance of a component is calculated by taking the reciprocal of its resistance. Siemens are 1/ohms.

Example 1.2 *Calculate the conductance of a conductor which has a resistance of 15 Ω.*

$G = 1/R = 1/15 = 0.067\,\text{S}$

1.2.6 Ohm's law

Because the use of Ohm's law is required in so many electrical calculations, it is important that its use is mastered. Fortunately, the law is very simple and therefore presents little difficulty provided careless slips are avoided. Ohm's law simply relates the voltage, V, across a resistance, R, carrying a current, I (Figure 1.5(a)).

Ohm's law states:

$$V = I \times R \qquad\qquad [1.3]$$

Example 1.3 *Calculate the current flowing through a 100 Ω resistor when connected across the terminals of a 12 V battery.*

$I = V/R = 12/100 = 0.12\,\text{A}$

Some people find the 'triangle method' of remembering Ohm's law useful; this is explained in Figure 1.5(b).

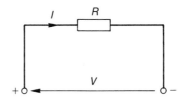

V, in volts, I in amps, R in ohms

$V = IR.$

Note: The voltage arrow, V, points in the increasingly positive direction.

(a)

Some people find the following triangle useful when applying Ohm's law:

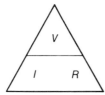

With one fingertip, cover the required quantity and see the appropriate formula:

Cover V see IR

Cover I see $\dfrac{V}{R}$

Cover R see $\dfrac{V}{I}$

Figure 1.5 *(a) Ohm's law; (b) aide-memoire*

(b)

1.2.7 The farad

The farad (unit symbol, F) is a measure of capacitance (C). The capacitance of a component, typically a pair of narrowly separated metal plates, is its ability to produce between the plates an electric field when a voltage is placed across the plates. This topic will be examined further in Section 1.3.2.

1.2.8 The henry

The henry (unit symbol, H) is a measure of inductance (L). The inductance of a component, usually a wire coil, is a measure of its ability to produce a magnetic field when carrying a current. We shall delve further into this topic in Section 1.3.3.

1.2.9 The watt

The watt (unit symbol, W) is a measure of electrical power dissipation (P). A watt is the power dissipated by a resistance carrying a current of one amp and having a voltage of one volt across it. Generally, watts = volts × amps:

$$P = V \times I \tag{1.4}$$

Example 1.4 *A 12 V battery is connected across a resistor and a current of 2 A flows. Calculate the power dissipated by the resistor.*

$P = V \times I = 12\,\text{V} \times 2\,\text{A} = 24\,\text{W}$

In Chapter 3 there are more explanations and examples of calculating power dissipation.

1.2.10 Hydraulic analogy

Students following a course of predominantly mechanical subjects often find that their understanding of the flow of electrical current is helped if it is compared to the flow of water through pipes. The comparison between the two systems is not always strictly true and some of the statements made below could be criticized by the purist. Nevertheless, they are broadly accurate. The analogy between electrical current flow and that of water is also explained by the diagrams and notes in Figure 1.6 and which support the following argument.

We know that the flow of conventional current actually comprises the movement of electrical particles in the opposite direction. The electrical particles (electrons) are made to move by an electrical force called an e.m.f. voltage). The particles are transported along a metal conductor (copper wire) because the electrical force at one end of the wire is greater

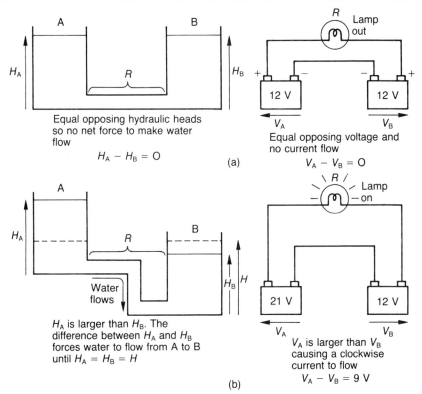

Figure 1.6 *Diagrams to show the similarity between the flow of electric current and the flow of water*

than that at the other end. In other words, there is a potential difference between the ends of the wire. Conventional current flows from the high force or potential to the low force or potential. The size and nature of the copper wire hinders (resistance) the movement of current through it. The electrons per second passing through the wire are called current, measured in amps, the pushing force is called the e.m.f., measured in volts and the wire's resistance to current flow is measured in ohms.

The hydraulic equivalent of the foregoing paragraph, with the words in brackets highlighting the particular electrical similarities, would read as follows:

'We know that the flow of water comprises the movement of many drips of water (electrons) all forming a continuous flow of the liquid (current). The water is forced to move by the influence of a hydraulic force, or water pressure, called a hydraulic head (volts). The water is transported through a metal pipe (copper wire) because the hydraulic head at one end of the pipe is higher than it is at the other end: there is a pressure difference (potential difference). Water (current) always flows from the high pressure (high potential) to the low pressure (low potential). The diameter of the pipe restricts the flow of water through it (electrical resistance). The flow of water is measured in litres per second (amps), the pushing force, causing the flow, is called the hydraulic head, measured in metres of water (volts) and the pipe's resistance to water flow is measured in loss of hydraulic head (not strictly equivalent to ohms!).'

1.2.11 Unit multipliers

Very often, the fundamental units mentioned above are inconveniently large or inconveniently small for a particular application in hand. For example, in electronic circuits, the currents involved are usually much less than one amp and therefore are more comfortably expressed as a number of milliamps (thousandths of an amp).

It is just the same as with our money system. We would say that something costs 2p rather than £0.02. However, to convert the 2p into the fundamental pound sterling unit, perhaps for accounting purposes, we would need to multiply by 0.01 because there are 100 pence in one pound. We could say that $2p = £2 \times 0.01$ but it is more usual to express the 0.01 as 10^{-2}. Therefore, $2p = £2 \times 10^{-2} = £0.02$ and the 10^{-2} term is called the multiplier.

Table 1.1 shows the most common multipliers used in electrical engineering:

It is useful to remember that $10^{x} = 1/10^{-x}$ and, conversely, that $10^{-x} = 1/10^{x}$.

Table 1.1

Name	Letter	Multiplier		
giga	G	10^{9}	or	1 000 000 000
mega	M	10^{6}	or	1 000 000
kilo	k	10^{3}	or	1000
centi	c	10^{-2}	or	0.01
milli	m	10^{-3}	or	0.001
micro	μ	10^{-6}	or	0.000 001
nano	n	10^{-9}	or	0.000 000 001
pico	p	10^{-12}	or	0.000 000 000 001

Example 1.5 *A current of 0.0146 A flows in a circuit. Convert this current flow into milliamps.*

To convert the 0.0146 A into mA we would need to divide the 0.0146 by 10^{-3} or 0.001. In either case the answer is 14.6 mA.

Example 1.6 *A current of 12 μA flows through a 5 k Ω resistor. Calculate the potential difference (p.d.) across the resistor and express this in mV.*

First use Ohm's law to establish the p.d. in volts, then convert this to mV:

$$V = IR = 12 \times 10^{-6} \times 5 \times 10^{3} = 60 \times 10^{-3} \text{ V}$$

Therefore, the p.d. across the resistor is 60 m V.

1.3 Electrical components

The Appendix shows a list of the component symbols used when drawing electrical and electronic circuits. This section is devoted to describing only the three most important passive components: the resistor, the capacitor and the inductor. Other electrical components such as the battery, generator, diode and transistor will be discussed in the appropriate chapters later in the book.

1.3.1 Resistors

Resistors are electrical components the purpose of which is to provide a restriction to the flow of electrons (current) through it. At first sight it may seem a little strange that we should wish to do this but there are, in fact, several useful side-effects. If we remember that a current passing through a resistor causes a volt drop across the resistor then its insertion can be used, for example:

- to limit the current in a circuit to prevent overheating
- to produce a voltage (across the added resistor) proportional to the circuit current.

Example 1.7   *A 24 W, 6 V lamp, having a filament resistance of 1.5 Ω, is to be connected to a 12 V battery. Calculate the value of resistor which must be placed in series with the 6 V lamp in order to prevent it from 'blowing'.*

From Equation 1.4, the current required to light the lamp properly is $24/6 = 4$ A. So, we need to add a resistor to the lamp input terminal such that 12 V can push only 4 A maximum through it. The total circuit resistance, comprising the lamp filament resistance of 1.5 Ω and the added current limiting resistor, R, can be calculated by Ohm's law as being 12 V/4 A $= 3$ Ω. Since the lamp contributes half of this resistance we need to add a further series resistor of $R = 1.5$ Ω. Figure 1.7 shows the circuit arrangement.

Example 1.7 raises another very important consideration when selecting a resistor to do a job: it must be of the correct power rating. In this particular case, our added series resistor of 1.5 Ω has to be capable of taking the full circuit current of 4 A without being damaged by heat. In fact it needs to be of the same power rating as the lamp it is protecting, namely 24 W. This is only to be expected since the resistor must itself pass a current of 4 A and drop across it 6 V of the battery's 12 V, leaving 6 V to drive the lamp correctly.

The value of a fixed resistor, measured in ohms, is decided in its manufacture. The type of material used, together with its physical dimen-

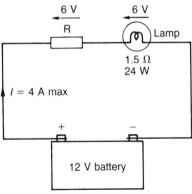

6 V   6 V

R   Lamp

1.5 Ω
24 W

I = 4 A max

+   −

12 V battery

Figure 1.7 *Using a resistor as a current limiter*

R must be 1.5 Ω to limit the
circuit current to 4 A

sions and method of construction, are the critical factors in deciding not only the ohmic value but also its power handing capability or power rating. The materials used are carbon or metal conducting films mounted on ceramic formers or high resistance wire wound around a non-conducting tube. Other resistors are manufactured to be readily variable in value and this is achieved either by hand or screwdriver adjustment. Figure 1.8 shows a few typical resistors that are used with electronic circuits.

The ohmic value of a resistor is usually marked on its body. The marking can be in the form of coloured bands or rings around the resistor or by printed numbers and letters. Figure 1.9 shows the resistor colour coding for resistors having a total of four coloured rings. The first two rings give the first two digits; the third ring indicates the number of zeros to be added and the fourth shows the manufacturing tolerance of the value.

Example 1.8 *A resistor is marked with the following coloured bands: brown, grey, black and silver. State the value and tolerance of the resistor.*

Band 1 brown   = 1
Band 2 grey   = 8
Band 3 black   = 0 (i.e. no zeros or unity multiplier)
Band 4 silver = ±10%

Resistance is 18 Ω ± 10%, that is, it must lie between 16.2 Ω and 19.8 Ω.

The method involving printing letters and numbers on the body of the resistor uses a letter instead of a decimal point and a second letter at the end to indicate the manufacturing tolerance. If the first letter is R or K or M it indicates ohms or kilohms or megohms respectively. If the final letter is F, G, J, K or M, a tolerance of ±1%, ±2%, ±5%, ±10% or ±20% respectively is indicated.

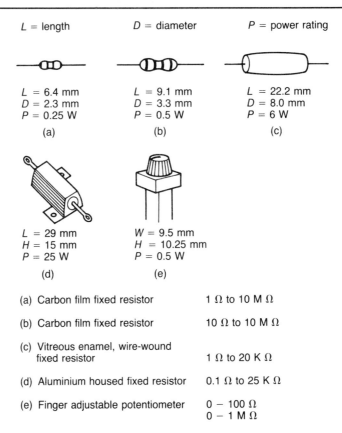

L = length	D = diameter	P = power rating

L = 6.4 mm
D = 2.3 mm
P = 0.25 W
(a)

L = 9.1 mm
D = 3.3 mm
P = 0.5 W
(b)

L = 22.2 mm
D = 8.0 mm
P = 6 W
(c)

L = 29 mm
H = 15 mm
P = 25 W
(d)

W = 9.5 mm
H = 10.25 mm
P = 0.5 W
(e)

(a) Carbon film fixed resistor   1 Ω to 10 M Ω

(b) Carbon film fixed resistor   10 Ω to 10 M Ω

(c) Vitreous enamel, wire-wound
    fixed resistor   1 Ω to 20 K Ω

(d) Aluminium housed fixed resistor   0.1 Ω to 25 K Ω

(e) Finger adjustable potentiometer   0 – 100 Ω
                                        0 – 1 M Ω

Figure 1.8 Typical resistor dimensions and ratings

Example 1.9   *Five resistors are marked with the following code: 18RK, 1R8J, R18G, 4K7K, 1M5K.*

$18RK = 18\,\Omega \pm 10\%$
$1R8J = 1.8\,\Omega \pm 5\%$
$R18G = 0.18\,\Omega \pm 2\%$
$4K7K = 4.7\,k\Omega \pm 10\%$
$1M5K = 1.5\,M\Omega \pm 10\%$

A final point on resistor values is that, like many items in the engineering world, resistors are manufactured for sale in 'preferred' values. Just as motor car tyres, hand spanners, nuts and bolts, and the like have to be selected for use from a range of standard sizes, then so it is with electronic components. The basic preferred values for resistors are as follows: 1.0, 1.2, 1.5, 1.8, 2.2, 2.7, 3.3, 4.7, 5.6, 6.8 and 8.2.

Example 1.10   *Calculations of the ideal resistor values required for a circuit are 0.9 Ω, 16 Ω, 155 Ω, 50 kΩ and 620 kΩ. State the nearest preferred values you would need to use.*

0.82 Ω, 15 Ω, 150 Ω, 47 kΩ and 680 kΩ respectively.

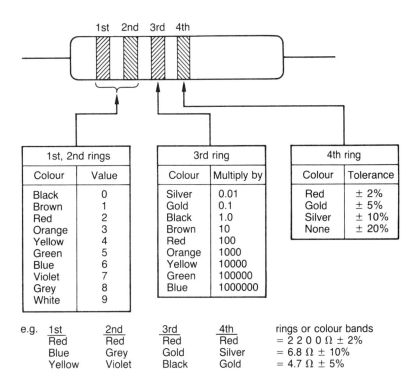

1st, 2nd rings	
Colour	Value
Black	0
Brown	1
Red	2
Orange	3
Yellow	4
Green	5
Blue	6
Violet	7
Grey	8
White	9

3rd ring	
Colour	Multiply by
Silver	0.01
Gold	0.1
Black	1.0
Brown	10
Red	100
Orange	1000
Yellow	10000
Green	100000
Blue	1000000

4th ring	
Colour	Tolerance
Red	± 2%
Gold	± 5%
Silver	± 10%
None	± 20%

e.g.

	1st	2nd	3rd	4th	rings or colour bands
	Red	Red	Red	Red	= 2 2 0 0 Ω ± 2%
	Blue	Grey	Gold	Silver	= 6.8 Ω ± 10%
	Yellow	Violet	Black	Gold	= 4.7 Ω ± 5%

Figure 1.9 *Resistor colour code for four-ring resistors*

1.3.2 Capacitors

Fundamentally, capacitors are nothing more than two metal plates spaced one from the other by a very small gap. If a positive potential is placed on one plate and a negative potential on the other, an electric field is set up between the plates across the gap. The capacitor is then said to be 'charged' because the electric field stores a form of energy which can be recovered to do useful work. The electric field can be regarded as 'lines of force' having a direction from the negative plate to the positive plate. Rather simplistically, these lines of force may be regarded as invisible rubber bands which can be pulled apart but which, when released, will spring back into place. If the space between the capacitor plates is filled, not with air but with mica, plastic, wax or other suitable material, the electric field storage capability can be much improved. The 'filling' material between the plates is called the capacitor dielectric. For a given size of capacitor, a large energy storage capacity is achieved if the dielectric is an aluminium-based electrolyte. The problem with these so-called electrolytic capacitors is that they must have a polarizing d.c. (direct current) applied to them otherwise the dielectric does not properly form and, further, the working voltage which can be applied across the plates is much reduced.

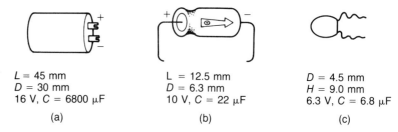

L = length, D = diameter, H = height
Voltages shown are recommended maximum
across terminals
C = capacitance at voltage shown.

Figure 1.10 *Typical capacitor dimensions and ratings. (a) Aluminium electrolytic – soldered tag connectors; (b) aluminium electrolytic – wire-ended axial; (c) solid tantalum*

L = 45 mm D = 30 mm 16 V, C = 6800 μF	L = 12.5 mm D = 6.3 mm 10 V, C = 22 μF	D = 4.5 mm H = 9.0 mm 6.3 V, C = 6.8 μF
(a)	(b)	(c)

Figure 1.10 shows three typical examples from the huge range of capacitors which are commercially available in all shapes and sizes. The value of the capacitor, in nanofarads (nF) or microfarads (μF), is usually marked on its body. The factors which decide the amount of capacitance offered by a particular capacitor are shown by the following equation:

$$C = \epsilon A/d \qquad\qquad\qquad [1.5]$$

where

C = capacitance in farads (F)
ϵ = dielectric constant (F/m)
A = plate area (m^2)
d = spacing between the plates (m).

In order to preserve space and yet use as large a plate area as possible, the flat plates, complete with the dielectric filling, are often rolled up into a cylindrical shape much in the manner of a 'Swiss-roll'.

Figure 1.11 shows the way the capacitor is made to store energy and how this stored energy can be recovered. Figure 1.11(a) shows the starting situation with the capacitor, C, uncharged and connected through an open switch, SW1, to a battery, B, and no current flows. Note that for the first time we have used the correct circuit symbol for the battery (see Appendix). Figure 1.11(b) shows the charging action which commences the moment SW1 is closed. The battery positive terminal then attracts the free electrons from the top layer of atoms in the capacitor upper plate. This movement of electrons constitutes a charging current flowing into the capacitor. The departure of electrons from the capacitor upper plate effectively leaves it positively charged because it now lacks its full complement of negative electrons. In a similar manner, the battery negative terminal sends electrons to lodge on the surface of the capacitor negative plate which, being surplus, effectively make it negatively charged. Thus, the flow of charging current, driven initially by the full battery voltage,

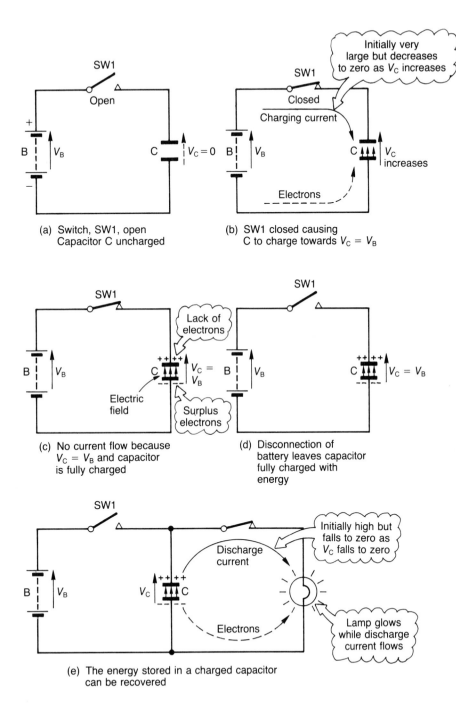

Figure 1.11 *Diagrams showing capacitor charge and discharge action*

V_B, is gradually reduced to zero as the capacitor voltage, V_C, builds up to oppose V_B. This latter situation is depicted by Figure 1.11(c). This shows the electric field between the capacitor plates as having been fully established. If the charging force, V_B is now removed by opening SW1, the capacitor still retains its electric field and the voltage across its plates remains at $V_C = V_B$. This is as shown by Figure 1.11(d). The addition of SW2 and a lamp load, as shown by Figure 1.11(e), indicates how the energy stored in the capacitor can be usefully used. The closing of SW2 will cause the lamp momentarily to light while the capacitor discharge current is flowing through it.

Figure 1.12 gives a graphical representation of the above action as controlled by the operation of SW1 and SW2. The application of V_B from time A does not immediately produce a fully charged capacitor. In fact, the capacitor takes its time in charging; t_{CH} to be precise, as shown in Figure 1.12(c). Similarly, the discharge of the capacitor through the lamp, initiated by the closing of SW2, takes time t_{DIS} to happen. This delay is true of all capacitor charging and discharging.

Understanding the action of capacitors in electronics is often helped if it is remembered that *the voltage between the plates of a capacitor cannot immediately be changed.*

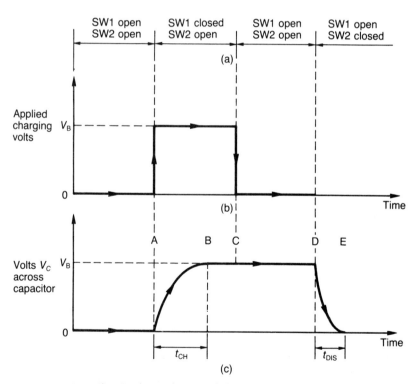

Figure 1.12 *Diagrams to show the delay in charging and discharging a capacitor*

t_{CH} = time to charge the capacitor
t_{DIS} = time to discharge the capacitor

This rule of not being able quickly to change the voltage across a capacitor is well illustrated by the following equation for calculating the instantaneous value of the current flowing into a capacitor:

$$i = C \frac{\delta v}{\delta t} \qquad [1.6]$$

where

i = instantaneous charging/discharging current (A)
C = the capacitance of the capacitor (F)
$\delta v / \delta t$ = the instantaneous rate of change of voltage across the capacitor (V/s).

Example 1.11 *A discharged, 1000 μF capacitor receives a constant charging current of 2mA for 0.5 s. Calculate the capacitor voltage at the end of the 0.5 s.*

i = constant 2 mA
C = 1000 μF
δt = 0.5 s.

Since the capacitor is originally discharged, its initial voltage will be zero so that the change of voltage, δv, will in fact be the final voltage of the capacitor after charging.

Substituting the known values in Equation 1.6 and solving for δv, we have:

$$\delta v = (i \times \delta t)/C = (2 \times 10^{-3} \times 0.5)/(1000 \times 10^{-6}) = 1.0\,\text{V}$$

We can see from Equation 1.6 that if we were *immediately* to change the voltage V_C by a small amount δv, it would require that δt be zero. This means that $\delta v / \delta t$ would be infinitely large (δv divided by zero!) and this would make the current i, infinitely large. The only way that the current could be infinitely large would be for it to be a short-circuit current. This is precisely what happens if a large voltage is suddenly applied to a capacitor. The capacitor plates immediately act as though they are welded together and pass a very high charging current. As the capacitor charges, the electric field between the plates starts to grow, the plates' voltages drift apart and the charging current gradually reduces to zero.

To summarize, *the voltage across the plates of a capacitor cannot be changed immediately.*

1.3.3 Inductors

The current flowing along any conductor causes a magnetic field to surround the length of the conductor. The field intensity is greatest at the surface of the conductor and gradually decreases as the distance from the conductor surface increases. The field can be imagined as being an invisible magnetic 'mist' around the conductor. The larger the current in the conductor the denser the 'mist' and the further it spreads away from the conductor.

Figure 1.13 shows the pattern of lines of magnetic force surrounding the current-carrying conductor. Note how the magnetic mist, or flux, is represented by lines tangential (parallel) to the surface of the conductor responsible for it. The lines of magnetic flux around a straight conductor are therefore represented by concentric circles. The stronger the magnetic field, the closer the circles are drawn together. The right-hand rule, Figures 1.13(a) and (b), and the corkscrew rule, Figures 1.13(c) and (d), show the relationship between the direction of flow of current and the direction of the resultant magnetic field.

By winding a length of wire into a solenoid comprising a coil of several loops, the current produces an increased magnetic flux. This effect is shown in progressive stages in Figure 1.14. The construction of two typical small inductors, as used in electronic circuits, is shown in Figure 1.15.

For a coil or solenoid having an air core and given dimensions, it is only the amount of current flowing through the coil that decides the intensity of the magnetic flux it produces. If the current flowing through

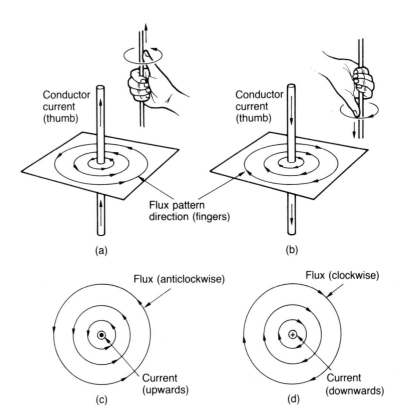

Figure 1.13 Magnetic field patterns surrounding a long current-carrying conductor. (a) and (c) current flowing upwards; (b) and (d) current flowing downwards.

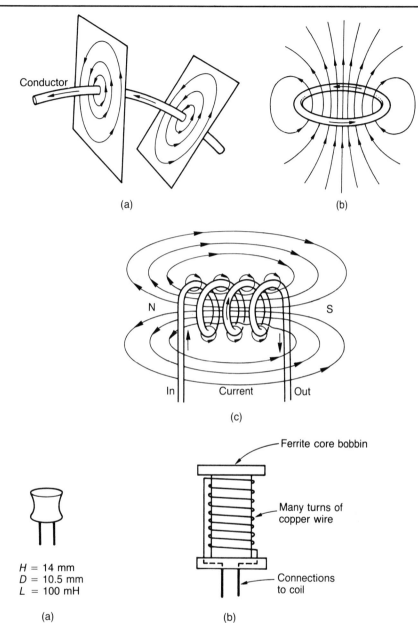

Figure 1.14 *Magnetic field patterns formed by currents flowing through (a) a curved conductor, (b) a loop conductor, (c) a solenoid*

Figure 1.15 *Typical fixed inductor construction. (a) Fixed inductor; (b) construction detail of fixed inductor*

$H = 14$ mm
$D = 10.5$ mm
$L = 100$ mH

the coil is increased from zero to *I* amps it will cause the associated magnetic field flux produced to increase from zero to Φ webers. We can say that a change of current δi has caused a change of flux $\delta \Phi$ and if there are *N* turns of conductor wire we find that for a particular design of coil $N \times \delta \Phi / \delta i$ is a constant, *L*. This constant is called the inductance of the coil and its basic unit is the henry:

$$L = N\frac{\delta\Phi}{\delta i}$$
[1.7]

While a current-carrying conductor produces a surrounding field of magnetic flux, Michael Faraday found also that when a conductor is made to pass through a magnetic field then an e.m.f. is set up across the length of the conductor. This induced e.m.f. causes a current to flow in the moving conductor. The magnitude of the induced e.m.f. is proportional to the speed at which the conductor cuts through the flux. We can write the rate of cutting flux as $\delta\Phi/\delta t$ webers per second. If this is multiplied by the number of coil turns, N, the product gives the value of e, the induced e.m.f.:

$$e = N \times \frac{\delta\Phi}{\delta t} \text{ volts}$$
[1.8]

Multiplying Equation 1.8 by $\delta i/\delta i$, that is, by 1, does not change its validity but enables us to rewrite the equation as:

$$e = N \times \frac{\delta\Phi}{\delta t} \times \frac{\delta i}{\delta i}$$

Rearranging this we obtain:

$$e = N \times \frac{\delta\Phi}{\delta i} \times \frac{\delta i}{\delta t}$$

From Equation 1.7 we know that $L = N\delta\Phi/\delta i$ so we can write:

$$e = L\frac{\delta i}{\delta t} \text{ volts}$$
[1.9]

From Equation 1.9 we can say that a coil having an inductance of one henry will have an e.m.f. of one volt developed across it if the current flowing through it changes at a rate of one amp per second.

The induced voltage, e, is in fact a back-e.m.f. in that it is in such a direction as to oppose the original voltage which is causing the current flow through the coil to change. For this reason, Equation 1.9 is often written with a negative value:

$$e = -L\frac{\delta i}{\delta t} \text{ volts}$$
[1.10]

Example 1.12   *The voltage across a circuit of negligible resistance is 1.5 millivolts when the current is falling at a rate of 600 microamps per second. Calculate the inductance of the circuit.*

The rate of change of current, $\delta i/\delta t = 600 \times 10^{-6}$ and the induced voltage, $e = 1.5 \times 10^{-3}$.

Transposing Equation 1.9 and substituting the above values we have:

$$L = (1.5 \times 10^{-3})/(600 \times 10^{-6}) = 2.5 \text{ H}$$

A final point before we leave inductors: Equation 1.10 shows that if we try to make $\delta i/\delta t$ too rapid (large) then the back-e.m.f. produced is sufficient to completely cancel the original driving voltage. The coil has then been effectively changed into a very high impedance and prevents rapid changes to the current flowing through it.

To summarize, *the current flowing through an inductance cannot be changed immediately.*

1.3.4 Transformers

A transformer is no more than a pair of coils which are so close together that the magnetic flux, Φ, produced by one coil fully links with the other. The circuit of the basic transformer is shown in Figure 1.16. If the flux, Φ, is generated by an alternating current in one coil (the primary winding) then the same alternating flux cutting the other coil (the secondary) induces a current in and a voltage across it. Because the flux is common to both windings, the primary and secondary coil voltages are related by the number of loop turns of wire, N_1 and N_2, forming the primary and secondary coils respectively. For example, if the flux links with only one turn in the primary but two turns in the secondary then the secondary voltage will be twice the primary voltage.

However, any increase (or step-up) in voltage caused by this transformer action must be paid for by there being an accompanying reduction in secondary current so as to keep the power input to the transformer equal to the power output from it.

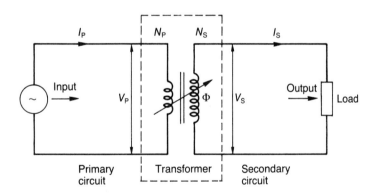

Φ is the alternating magnetic flux linking N_P and N_S
I_P, V_P, I_S and V_S are a.c. currents and voltages.

Power input = power output

$\therefore V_P I_P = V_S I_S$

$$V_S = \frac{N_S}{N_P} \times V_P$$

Figure 1.16 *Basic transformer*

$$I_S = \frac{N_P}{N_S} \times I_P$$

Example 1.13   *A transformer has a primary winding of 100 turns and a secondary of 2000 turns. The primary winding is connected across a 240 V, 50 Hz mains supply and draws from it a current of 1 A. Calculate the secondary voltage and current assuming the transformer to be 100% efficient.*

The secondary winding has the larger number of turns so the common flux will produce a larger voltage across the secondary than it will across the primary:

$V_S = V_P \times N_S/N_P = 240V \times 2000/100 = 4800\,V$

The 100% efficient transformer has equal input and output powers so we can say:

$I_S = I_P \times V_P/V_S = 1\,A \times 100/2000 = 0.05\,A$

2 Power sources

2.1 Electrical cells and batteries

An electrical cell comprises a container holding a liquid or paste, called an electrolyte, into which two dissimilar metal rods or plates, called electrodes, are placed some distance apart. If the two electrodes are connected together externally by a conducting wire, an electric current circulates from one electrode along the wire to the other electrode and back through the electrolyte. An e.m.f. or voltage is set up between the electrodes and it is this chemically generated force which drives the current around the circuit.

Unfortunately, the passage of current through the electrodes and the electrolyte causes chemical damage and after a while the current flow ceases. However, the materials used in the make-up of some cells allows the process to be reversed. If a current is made to flow in the reverse direction the cell electrodes and electrolyte revert to their original states. The cells which permit this reverse processing, or recharging, are usually more expensive than those in which the chemical action is non-reversible. The non-rechargeable cells are called *primary cells* and the rechargeable ones are called *secondary cells*.

So, an electrical cell is a means of storing electrical energy within the chemicals it contains. A battery is made by connecting two or more cells together to produce an increased voltage output.

2.1.1 Ampere-hour (Ah) capacity

The product of the current, in amps, drawn from a battery and the time, in hours, over which it is drawn is a measure of the amount of energy stored in a battery, in Ah. In general, if a battery is given a storage rating of, say, 50 Ah, it means that the fully charged battery can produce a steady flow of 1 amp for 50 hours or 10 amps for 5 hours or 5 amps for 10 hours and so on. In practice it is not quite that simple, but for our purposes it is good enough.

A simple way of looking at a battery is to regard it as being packed with electrons (i.e. current) held at a certain pressure (voltage). If the battery is connected to a load circuit, say a lamp, the battery voltage pushes the stored electrons through the lamp and the lamp lights. The current will continue to flow through the lamp so long as the battery can maintain its chemical action and hence its rated voltage across its output terminals. If a motor car 12 V battery is rated at 38 Ah, then with all the vehicle lights switched on and the engine switched off the battery could be expected to maintain a typical total load of 120 W for 3.8 hours.

2.1.2 Primary and secondary cells and batteries

Figure 2.1 shows the arrangement for the popular primary cell used in hand torches, wall clocks and the like. The cheapest is the zinc–carbon cell. The carbon rod and the zinc container act as the metal electrodes and the electrolyte is a paste used with a polarizing material which enhances its action. The cell produces about 1.5 V and typically has a 0.05 to 1.0 Ah capacity. The zinc–carbon cell is non-rechargeable and when exhausted must be replaced by a new one. When exhausted it is the zinc container which has been badly eroded and, unless the cell is of the sealed type, this may allow the electrolyte paste to leak and cause damage.

In many instances, the cost of replacing exhausted primary batteries in a modern torch or transistorized radio represents a large proportion of

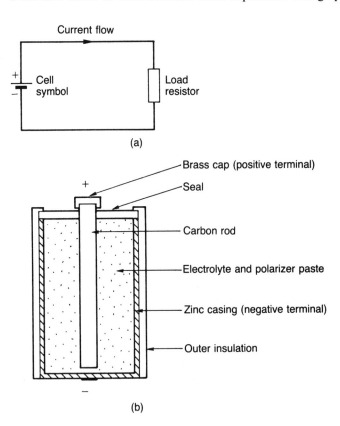

Figure 2.1 *(a) Single cell connected to a load; (b) construction of zinc–carbon primary cell*

the original cost of the equipment. It is these economics that have resulted in the development of a range of rechargeable secondary cells which are interchangeable with existing primary cells. These rechargeable cells typically use nickel–cadmium electrodes and have alkaline electrolytes. They are more expensive than the equivalent zinc–carbon cell and need a battery charger. Figures 2.2 and 2.3 show a range of these cells as used in small electrical and electronic equipment.

For heavier duty, the d.c. power source is usually a large rechargeable secondary lead–acid battery using a number of cells each of which produces around 2.1 V when fully charged. From three to thirty-six such cells provide a nominal battery voltage of between 6 V and 72 V. Figure 2.4 shows the principle of construction of this type of cell which is typically connected in a series of six to produce the 12 V battery used by most motor cars. The most modern of these batteries are now sealed for life and require minimal maintenance. Figure 2.5 shows a range of small sealed lead–acid batteries. Much larger secondary batteries weighing several tonnes are used in forklift trucks and milk floats.

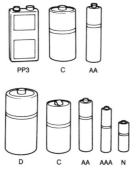

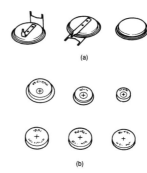

Figure 2.2 *A range of rechargeable nickel–cadmium cells and batteries of equivalent size to the popular non-rechargeable zinc–carbon primary types*

Figure 2.3 *(a) Non-rechargeable coin cells suitable for mounting on electronic printed circuit boards; (b) non-rechargeable button cells suitable for use in calculators, cameras, clocks and watches*

2.2 Power supply units (PSUs)

As an alternative to using primary and secondary batteries as d.c. power sources, it is often more convenient to have an a.c. mains-driven PSU. This is essentially plugged into the mains supply at one end and a choice of d.c. or a.c. voltages are available at the other end. Figure 2.6 shows the simplest PSU, the battery charger. The transformer reduces the 240 V a.c. mains input to about 20 V a.c., the rectifier unit changes the a.c. into d.c. pulses and the smoothing capacitor produces a near steady d.c. of around 14 V which is suitable for forcing a reverse charging current through the nominal 12 V battery. Figure 2.7(a) shows a simple lead-acid battery charger and Figure 2.7(b) an automatic charger suitable for the smaller nickel–cadmium type.

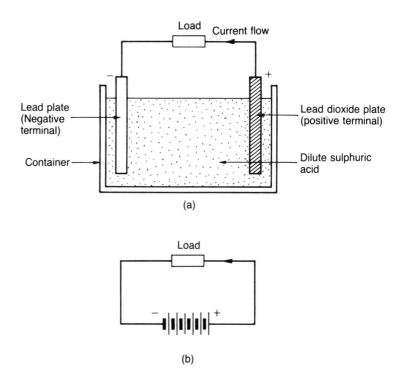

Figure 2.4 *(a) Principle of a single lead–acid secondary cell; (b) six cells forming a battery*

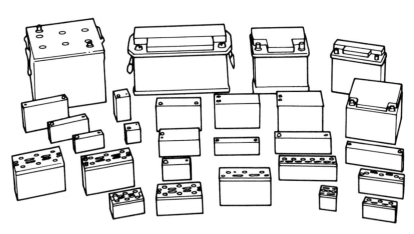

Figure 2.5 *A range of sealed lead–acid rechargeable batteries ranging from 2 V to 12 V d.c. and having a storage capacity ranging between 1.0 Ah and 110 Ah*

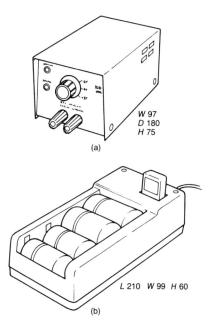

Figure 2.6 *Block diagram of a simple power supply used to recharge a battery*

Reduced a.c. voltage

Rough d.c.

Smooth d.c.

a.c. → mains input

Mains step-down transformer

Rectifier unit

Smoothing capacitor

12 V battery being charged

Figure 2.7 *Power supply units used as battery chargers. (a) A simple 2 V to 12 V adjustable charger for the lead–acid batteries shown in Figure 2.5; (b) an automatic charger unit which will accommodate a range of small cells as shown in Figure 2.2*

W 97
D 180
H 75

(a)

L 210 W 99 H 60

(b)

2.3 Generation of single-phase currents

2.3.1 How a simple single-phase alternator works

The principle involved in a single-phase alternator (Figure 2.8) is based on the findings of Michael Faraday. If a conductor is moved through a magnetic field flux an e.m.f. is induced between the ends of the conductor. This causes an induced current to flow along the conductor. Faraday further found that the faster the conductor is moved, the greater the induced e.m.f. and its associated current. Not only does the magnitude of the induced e.m.f. depend upon the rate at which the conductor cuts the magnetic flux, the density of the flux and the length of the moving conductor also have an effect. Since the maximum rate of cutting flux

E.m.f. between A and B = e_{AB} = Bv/ sin θ volts.

Rotating conductor loop, ABCD, is shown in
the position where θ = 90°

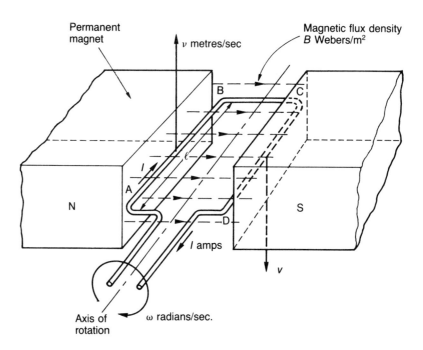

Figure 2.8 *Principle of single
loop, single-phase a.c. generator
(alternator)*

will occur if the conductor passes through it at right-angles, the angle of
flux cutting is also important.

If we call the flux density B webers/m², or tesla, the length of the
conductor in the magnetic flux l m, the speed of the conductor passing
through the flux v m/s, and the angle θ, then the induced e.m.f., e volts,
across the length l, is given by the equation:

$$e = Bvl \sin \theta \text{ volts} \hspace{4cm} [2.1]$$

The magnetic flux is produced by the permanent magnet with the
magnetic flux lines conventionally regarded as having a direction from
the N pole to the S pole. Figure 2.8 effectively uses two conductors, AB
and DC, which, by being the two long sides of a continuous loop, are
connected in series. The loop is mechanically rotated by some external
means. The loop is made to rotate at a constant angular velocity, ω rad/s,
and this gives the two conductors a linear velocity v m/s. Conductor AB
moves upwards, causing an induced current to flow along it from A to B.
Conductor CD moves downwards, causing an induced current flow in
the opposite direction from C to D. The loop end conductors, AD and
BC, are regarded as being outside the magnetic field and have no induced
e.m.f. across them. Each conductor produces its own induced e.m.f. and

with the two being in series the total output from the ends A and D is $2 \times e$ volts.

Figure 2.9(a) shows the effect of the circular rotation of the loop. In the horizontal position, the conductors AB and CD are cutting the flux at the maximum rate, that is, at right angles with $\theta = 90°$. However, a further 90° rotation to make $\theta = 180°$ results in the flux lines being tangential to the conductor movement; there is no cutting of flux and no induced e.m.f. The complete circular movement of the conductors produces two points where the induced e.m.f. is zero (0° and 180°) and two points of maximum induced e.m.f. (90° and 270°). Figure 2.9(b) shows the resulting output voltage waveform from the terminals A and D to be a sine wave given by $e_{AD} = 2Bvl \sin \theta$ volts. Now B and l are

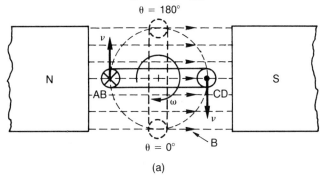

(a)

Voltage output from open loop ends, A→D is the sum of the two e.m.f.s induced into conductors AB and CD.

$e_{AD} = 2 Bv\ell \sin \theta$ volts

Since $2Bv\ell$ is constant and is the maximum value of e_{AD} when $|\sin \theta| = 1$, i.e. $\theta = 90°$ or 27ᴟ , then we say $2Bv\ell = V_{max}$ and we can write the loop output voltage as:

$e_{AD} = V_{max} \sin \theta$

Therefore we say that e_{AD} is a sine wave or that the output voltage, e_{AD}, varies sinusoidally.

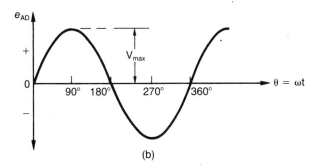

Figure 2.9 *The rotating loop conductor in a fixed magnetic field (a) produces a sinusoidal voltage output (b)*

(b)

decided by the manufacturer of the alternator and if the loop conductor is rotated at a constant speed, the term $2Bvl$ is constant and can be written as V_{max} because the maximum value of e_{AD} is indeed $2Bvl$ occurring when $\sin\theta = 1$, i.e. when θ is 90° and 270°.

Therefore we can express the sinusoidal output voltage from a single-phase alternator by the general term $V_{max}\sin\theta$. Also, since ω is constant we can replace θ by ωt and Equation 2.1 for the instantaneous output voltage of the alternator becomes:

$$v = V_{max}\sin\omega t \text{ volts} \qquad [2.2]$$

2.3.2 The practical single-phase alternator

Figure 2.10 shows a more practical single-phase alternator. The conductors into which the generated current is induced comprise more than the single loop shown in Figure 2.8 and, in practice, are wound as stationary field windings around a pair of pole pieces. The rotor in this case is shown as a permanent magnet (but it could be an electromagnet) which produces a rotating magnetic field. The stator field winding, F, receives an induced sinusoidal e.m.f. as it is swept by magnetic flux from the passing rotor N pole. At the same time the stator winding, F_1, is being similarly affected by the rotor S pole. The two e.m.f.s from windings F and F_1 are added together to provide the final sinusoidal voltage output.

2.4 Generation of three-phase currents

Figure 2.11(a) is a somewhat simplistic diagram to show how, in order to produce three single-phase output voltages, the single-phase alternator of

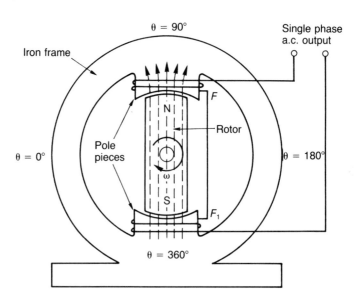

Figure 2.10 *Practical single-phase alternator. The rotating permanent magnet provides a rotating magnetic field which interacts with the stationary field coils, F and F_1, connected to produce an a.c. output*

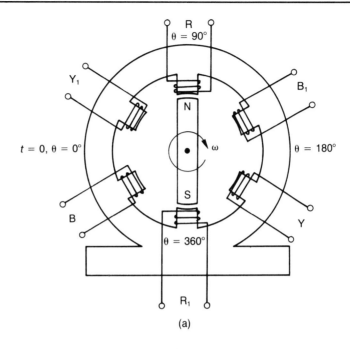

(a)

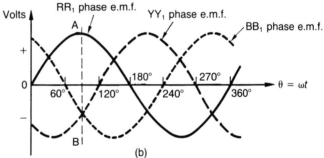

(b)

Line AB indicates stator coil (phase)
e.m.f.s for the rotor position shown in (a)

The stator (phase) coils RR$_1$, YY$_1$ and BB$_1$ are interconnected
in such a way that when the N pole of the rotating
magnet (rotor) is beneath R, Y and B a positive-phase
e.m.f. is induced. The N pole interacts with phase coils
R$_1$, Y$_1$ and B$_1$ to induce a negative-phase e.m.f.

Figure 2.11 *Three-phase
alternator. (a) Windings
arrangement; (b) phase e.m.f.s*

Figure 2.10 simply requires the addition of another two pairs of series-
aiding stator windings. The three pairs of stator windings, RR$_1$, YY$_1$ and
BB$_1$, are positioned 120° apart. The magnetic flux from the rotating
permanent magnet sweeps each pair in that order and three separate
sinusoidal *phase* voltages, each 120° lagging the other, are obtained.
Figure 2.11(b) shows a graphical representation of the three-phase voltage waveforms. The phases are sometimes labelled as phases 1, 2 and 3

in the time order that they reach their positive maximum values; equally, they are also known as the red, yellow and blue phases. The associated wiring insulation is coloured accordingly.

Figure 2.12(a) shows the conventional electrical diagram of the stator output windings. The rotor and its rotating magnetic field are not shown and, for this particular diagram, it is assumed to be rotating clockwise in the phase order R, Y, B. The six wires shown as being necessary to carry current from the three phases can be reduced to four by the connection of R_1, Y_1 and B_1 into a single conductor to form what is known as the *neutral* line.

The three-phase alternator is constructed using identical phase windings and these are connected to three loads. Just as it is normally good engineering practice to load each of the four or six cylinders of a motor vehicle equally, then so it is with the three phases of a three-phase alternator. In fact, it can be an important part of an electrical engineer's

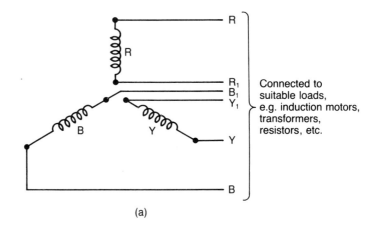

(a)

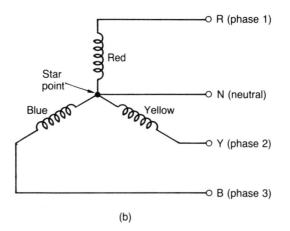

Figure 2.12 *(a) The three separate phase windings can be shown in an electrical diagram. (b) If the ends R_1, Y_1 and B_1 are connected to form a single wire, called the neutral, a three-phase four-wire system is produced*

(b)

responsibility to ensure that the three phases are *balanced*. A three-phase system is said to have a balanced load when the current drawn from each of the three phases is the same. When this happens the neutral line, which carries any out-of-balance currents, can be omitted leaving the cost of only three supply wires. The matter of balanced loads is covered in more detail in Chapter 5.

2.5 Generation of direct current

This is achieved by a single-phase alternator fitted with a segmented *commutator* in contact with a pair of stationary carbon pick-up brushes. This situation is shown by Figure 2.13(a) which, for simplicity, comprises only a single loop and a simple two-segment commutator. In practice, there would be eight or more separate loops, each with its own pair of

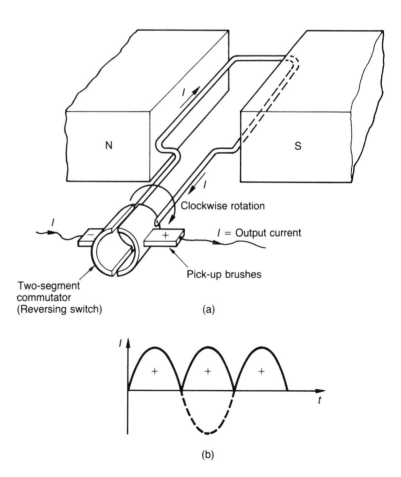

Figure 2.13 Principle of the single-loop d.c. generator. (a) Commutator action produces unidirectional current output; (b) unidirectional output current pulses

commutating segments, and angularly spaced to form a cylindrical, rotating *armature*.

The action of the commutator is to produce a unidirectional output current from the output pick-up brushes. This is achieved, despite the mechanical rotation of the loop and its commutator, by the positive output brush always being connected, through the appropriate commutator segment, to the right-hand loop conductor. The unidirectional d.c. output waveform really comprises a series of positive-going half sine waves as shown by Figure 2.13(b). The greater the number of commutator segments and associated armature loop conductors, the nearer the output waveform approximates to a true d.c.

Exercises

2.1   A motor car is fitted with a 12 V, 44 Ah battery. If the car is parked overnight with its two 48 W headlights and two 7 W tail-lights left on, calculate the time taken to completely discharge the battery.

2.2   A motor car has a 12 V, 38 Ah battery and a 1200 W starter motor. Calculate the total engine cranking time available without the battery being recharged.

2.3   A straight 200 mm long conductor moves at right angles through a magnetic field of 1.2 tesla at a speed of 30 m/s. Calculate the voltage induced across the conductor.

2.4   If in question 2.3 the conductor moved through the magnetic field at 45° to the normal at an increased speed of 60 m/s, calculate the new voltage generated.

2.5   A straight axial conductor 30 mm long is fixed on the surface of a cylindrical armature of diameter 45 mm. If the armature rotates at 2000 rev/min and the magnetic flux density normal to the conductor motion is 0.65 tesla, calculate the e.m.f. induced in the conductor.

2.6   If in question 2.5 the speed of rotation were halved and the flux density reduced to 0.6 tesla, calculate the new e.m.f. induced in the conductor.

3 Direct current network analyses

3.1 Series networks

3.1.1 Voltage

If we imagine a potential difference, V volts, as the electrical force which pushes a current, I amps, along a wire having a resistance to current flow of R ohms, we can connect the three quantities by Ohm's law:

$$V = IR$$

or

Volts = amps × ohms

or

Voltage = current × resistance

Example 3.1 *If a motor car battery is connected across a resistance of 1.5 Ω, calculate the current which flows through the resistance.*

The problem can be represented by either the full circuit in Figure 3.1(a) or by Figure 3.1(b). Because there is no other resistance or other component in the circuit, the full battery voltage is applied across the 1.5 Ω resistor. Clearly, the 12 V battery provides the driving force to push the current through the resistance of 1.5 Ω. In doing this, the full 12 V of the battery are 'dropped' across the resistor. The circuit current, I_S, supplied by the supply voltage, V_S, is given by:

$$I_S = \frac{V_S}{R}$$
$$= \frac{12\,\text{V}}{1.5\,\Omega}$$
$$= 8\,\text{A}$$

Just the same principle is involved in the motor car cooling system shown by Figure 3.2. The water pump provides the driving pressure (battery) to push the liquid coolant (current) through the hot engine and the cooling radiator (resistance). The 'volt drop', or reduction of

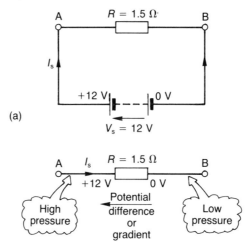

Figure 3.1 *(a) Simple series circuit to illustrate Ohm's law. (b) Voltage can be regarded as a force or pressure*

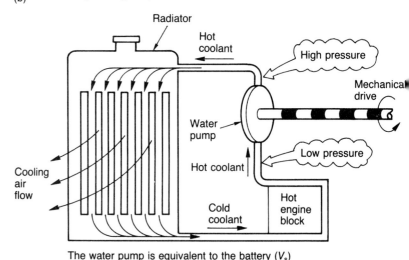

Figure 3.2 *The operation of the simple electrical circuit in Figure 3.1 is much the same as that of a motor car cooling system*

electrical pressure from one end to the other, across any electrical component is necessary in order to force the current through the component. All electrical components have a natural unwillingness to allow current to flow through them.

Should we wish to know the value of the current flowing through a circuit in which there are two or more components in *series*, we simply add together the different resistance values and treat the sum as a single resistor.

Example 3.2   *For the circuit shown in Figure 3.3, calculate the total resistance of the circuit and hence the circuit current supplied by the battery.*

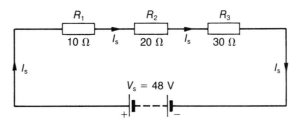

Figure 3.3 *Three resistors in series*

The total resistance is:

$$R_T = R_1 + R_2 + R_3$$
$$= 10\ \Omega + 20\ \Omega + 30\ \Omega$$
$$= 60\ \Omega$$

The supply current, I_S, is regarded by convention as flowing out of the battery positive terminal, through each resistor and returning to the battery negative terminal. Note that there is no loss in the amount of current (amps) flowing around the circuit, just as there is no loss in coolant flowing around the motor car cooling system. In other words, all of the current that leaves the battery returns to it. The constant value of I_S is given by

$$I_S = \frac{V_S}{R_T} = \frac{48\ \text{V}}{60\ \Omega} = 0.8\ \text{A}$$

What is lost, as the current flows around the circuit from the battery positive to the battery negative terminal, is the electrical pressure (voltage). In order to investigate this loss of electrical pressure (volt drop) we can redraw Figure 3.3 in the slightly different form shown in Figure 3.4.

The same three series resistors and the current source (the battery) are now shown connected together by the circuit conducting wire which has been drawn so as to indicate the different sections of like voltage, V_1, V_2, V_3 and V_4.

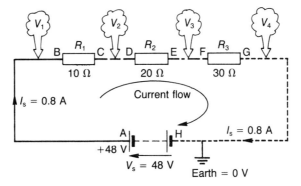

Figure 3.4 *The same current flows through each resistor and the battery. Different sections of the circuit are at different potentials or voltages*

We can calculate the respective values of V_1 to V_4 by finding the volt drop across each resistor caused by the common circuit current flowing through it, and progressively deducting these drops from the full battery voltage applied to the circuit. It often makes life easier if the negative terminal of the battery is regarded as being at zero or earth potential and then all other voltages are regarded as being positive with respect to earth.

To find the value of V_1

This is simply the voltage of the conductor section A to B. Since we regard all conductors as being perfect, that is, having a zero resistance, there can be no volt drop between A and B. Therefore, V_1 will be the same voltage as the battery positive terminal, namely:

$$V_1 = V_S = 48 \text{ V}$$

To find the value of V_2

This is the voltage of the conductor section B to C. V_2 will be V_1 less the volt drop across R_1 caused by the flow of I_S through R_1. The volt drop across R_1 by Ohm's law is:

$$R_1 \times I_S = 10 \ \Omega \times 0.8 \text{ A} = 8.0 \text{ V}$$

Therefore, $V_2 = V_1 - 8.0 \text{ V} = 48.0 \text{ V} - 8.0 \text{ V} = 40 \text{ V}$.

To find the value of V_3

This is the voltage of the conductor section E to F:

$$
\begin{aligned}
V_3 &= V_2 - \text{volt drop across } R_2 \\
&= V_2 - I_S R_2 = 40.0 \text{ V} - 0.8 \text{ A} \times 20 \ \Omega \\
&= 40.0 \text{ V} - 16.0 \text{ V} \\
&= 24 \text{ V}
\end{aligned}
$$

To find the value of V_4

This is the voltage of the section G to H:

$$
\begin{aligned}
V_4 &= V_3 - \text{volt drop across } R_3 \\
&= V_3 - I_S R_3 = 24.0 \text{ V} - 0.8 \text{ A} \times 30 \ \Omega \\
&= 24.0 \text{ V} - 24.0 \text{ V} \\
&= 0 \text{ V}
\end{aligned}
$$

Alternatively, because the conductor section G to H is connected directly to the battery, V_4 must be the same voltage as the battery negative terminal. This battery terminal is earthed to zero volts so V_4 must likewise be clamped to zero volts potential.

Note that the total applied external voltage from the battery is 48 V and that this is exactly equal to the sum of the three volt drops across the three resistors.

Example 3.3   *If in Figure 3.4 the values of R_1, R_2 and R_3 are 5, 10 and 15 Ω respectively, calculate the circuit current flowing and the volt drop across each of the resistors if the battery voltage is 24 V.*

The total circuit resistance, $R_T = 5\ \Omega + 10\ \Omega + 15\ \Omega = 40\ \Omega$.
The circuit current, $I_S = V_S/R_T = 24\ \text{V}/40\ \Omega = 0.6\ \text{A}$.
Let the volt drop across resistor R_1 be V_{R1} and so on.

$$V_{R1} = \text{current through } R_1 \times \text{resistance of } R_1$$
$$= \qquad 0.6\,\text{A} \qquad \times \qquad 5\ \Omega$$
$$= 3\,\text{V}$$

$$V_{R2} = I_S R_2$$
$$= 0.6\,\text{A} \times 15\ \Omega$$
$$= 9\ \text{V}$$

$$V_{R3} = I_S R_3$$
$$= 0.6\,\text{A} \times 20\ \Omega$$
$$= 12\,\text{V}$$

Notice that we can quickly check our working by summing V_{R1}, V_{R2} and V_{R3} and comparing the result with the battery voltage. In this case, $3 + 9 + 12 = 24\,\text{V}$ so all is well.

3.1.2 Power dissipated

The power dissipated in any circuit is always in the resistance elements. The unit of power is the *watt* and it is the product of the amps and volts flowing through and across a resistance respectively. *Be careful to use the volt drop across the resistance under consideration* and *not* the supply voltage, V_S, unless they happen to be the same.

Example 3.4   *A 12 V battery supplies current to a lamp having a filament coil resistance of 3 Ω. Calculate the power dissipated by the lamp.*

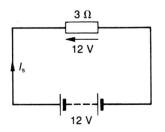

Figure 3.5 *Circuit for Example 3.4*

The circuit is shown in Figure 3.5. The full battery voltage is developed across the single 3 Ω resistor so we can calculate the current, I_S, flowing through the resistor:

$$I_S = \frac{\text{volts across resistance}}{\text{resistance in ohms}}$$
$$= \frac{12\,\text{V}}{3\,\Omega} = 4\,\text{A}$$

Therefore, the power dissipated by the lamp = (the current through the lamp) × (the volt drop across the lamp):

$$= 4\,\text{A} \times 12\,\text{V} = 48\,\text{W}$$

This is typically the power rating of a motor car headlamp.

Now let us try the power calculation for a circuit having two components.

Example 3.5   *The circuit in Figure 3.6 is for a battery supplying current to two 3 Ω filament lamps in series. Calculate the power dissipated by each lamp and the total power dissipation.*

Once again we start to solve the problem by calculating the total circuit resistance. This is then used in Ohm's law, together with the battery or other source voltage in order to find the circuit current flowing.

The total circuit resistance is 3 Ω + 3 Ω = 6 Ω

The circuit current is 12 V/6 Ω = 2 A

The 2 A current flows through both of the 3 Ω resistors and so to find the power dissipated by each resistor we must find the volt drop across each of them. Clearly, since the two resistors are of equal ohmic value (3 Ω) they must equally share the 12 V battery voltage. Therefore, each resistor must drop 6 V across it. Alternatively, this could have been determined by multiplying the 3 Ω resistance by the current flowing through it: 3 Ω × 2 A = 6 V.

The power dissipated by a 3 Ω resistor carrying a current of 2 A and having a 6 V drop across it is:

$$6\,\text{V} \times 2\,\text{A} = 12\,\text{W}$$

The total power dissipation by two identical resistors is 24 W.

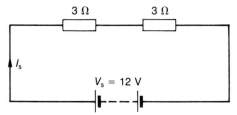

Figure 3.6 *Circuit for Example 3.5*

So, if we have a resistor of R ohms carrying a current of I amps and having a volt drop across it of V volts, we can say that the power dissipated by the resistor is W watts:

$$W = V \times I \qquad\qquad [3.1]$$

But, by Ohm's law, $V = IR$ and so in Equation 3.1 we can replace V by IR. This gives $W = (IR)I$:

$$W = I^2 R \qquad\qquad [3.2]$$

Similarly, since $I = V/R$, we can replace I in Equation 3.1 by V/R. This gives $W = V(V/R)$:

$$W = \frac{V^2}{R} \qquad\qquad [3.3]$$

These three equations should be memorized so that if any pair of V, I and R is known, the power dissipated can readily be calculated.

Example 3.6   *In the circuit shown in Figure 3.7, calculate the power dissipated by each resistor and the battery voltage.*

Power in $R_{2\Omega} = I^2 R = (2A)^2 \times 2\,\Omega = 8\,\text{W}$
Power in $R_{3\Omega} = I^2 R = (2A)^2 \times 3\,\Omega = 12\,\text{W}$
Power in $R_{4\Omega} = I^2 R = (2A)^2 \times 4\,\Omega = 16\,\text{W}$
Power in $R_{5\Omega} = I^2 R = (2A)^2 \times 5\,\Omega = 20\,\text{W}$

Notice that we can calculate the battery voltage, V_1, by again using Ohm's law:

$$
\begin{aligned}
V &= \text{(circuit current)} \times \text{(total circuit resistance)} \\
  &= 2\,\text{A} \times (2 + 3 + 4 + 5)\,\Omega \\
  &= 2\,\text{A} \times 14\,\Omega \\
  &= 28\,\text{V}
\end{aligned}
$$

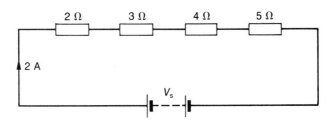

Figure 3.7 *Circuit for Example 3.6*

The total circuit power dissipation is:

(circuit current)^2 × (total circuit resistance)

$$= (2A)^2 \qquad \times 14\,\Omega$$
$$= 56\,\text{W}$$

or

(circuit current) × (battery voltage)
$$= 2\,\text{A} \qquad \times 28\,\text{V}$$

or

(battery voltage)^2 × (total circuit resistance)

$$= (28\,\text{V})^2/14\,\Omega$$
$$= 56\,\text{W}$$

3.2 Parallel networks

Suppose we connect these resistors in parallel across a battery of voltage V_S. Figure 3.8(a) shows the circuit. The battery in fact 'sees' the three parallel resistors effectively as a single resistor which we can call R_E. This is shown in Figure 3.8(b).

In the previous circuits where all the resistors were in series, the same circuit current, the supply current, I_S, flowed through all of them. In the parallel circuit, the supply current is not common to all three resistors. In fact, the values of the branch resistors R_1, R_2 and R_3 decide the individual branch currents I_1, I_2 and I_3. Clearly, the number of electrons making up the supply current, I_S, is divided between the three branches in inverse proportion to the resistances of the branches. The quantity

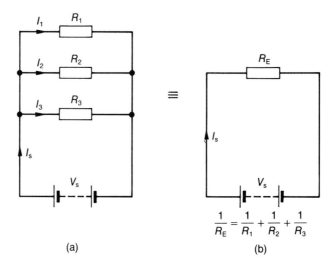

Figure 3.8 R_E *is the equivalent single resistor of* R_1, R_2 *and* R_3 *in parallel*

(a)   (b)

which is common to all branches in the parallel circuit is the supply voltage, V_S, across them.

3.2.1 Current flowing through each parallel branch

The current, I_1, flowing through the branch containing R_1 is calculated by Ohm's law:

$$I_1 = \frac{V_S}{R_1}$$

Similarly, the currents in the other two branches are:

$$I_2 = \frac{V_S}{R_2} \quad \text{and} \quad I_3 = \frac{V_S}{R_3}$$

Now, since we know that $I_S = I_1 + I_2 + I_3$, we can write:

$$I_S = \frac{V_S}{R_1} + \frac{V_S}{R_2} + \frac{V_S}{R_3} \qquad [3.4]$$

But from Figure 3.8(b) we can see that I_S is given by $I_S = V_S/R_E$, so we can rewrite Equation 3.4 as:

$$\frac{V_S}{R_E} = \frac{V_S}{R_1} + \frac{V_S}{R_2} + \frac{V_S}{R_3}$$

Cancelling the common factor, V_S, on both sides of the equals sign, we have:

$$\frac{1}{R_E} = \frac{1}{R_1} + \frac{1}{R_2} + \frac{1}{R_3}$$

For N parallel resistors, as shown in Figure 3.9, the equivalent single resistor, R_E, can be calculated from:

$$\frac{1}{R_E} = \frac{1}{R_1} + \frac{1}{R_2} + \frac{1}{R_3} + \frac{1}{R_4} + \ldots + \frac{1}{R_N} \qquad [3.5]$$

For the case where there are two (*and only two!*) parallel resistors, Equation 3.5 becomes:

$$\frac{1}{R_E} = \frac{1}{R_1} + \frac{1}{R_2}$$

So

$$\frac{1}{R_E} = \frac{R_1 + R_2}{R_1 R_2}$$

and inverting both sides gives:

$$R_E = \frac{R_1 R_2}{R_1 + R_2} = \frac{\text{product}}{\text{sum}} \qquad [3.6]$$

This is known as the *product-over-sum relationship* for quickly calculating the equivalent single resistance for *two resistors* in parallel.

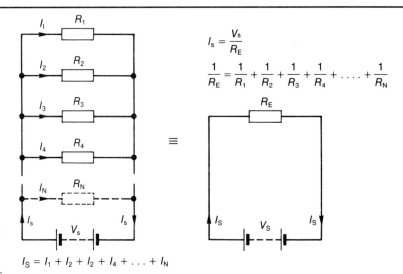

$$I_s = \frac{V_s}{R_E}$$

$$\frac{1}{R_E} = \frac{1}{R_1} + \frac{1}{R_2} + \frac{1}{R_3} + \frac{1}{R_4} + \dots + \frac{1}{R_N}$$

Figure 3.9 *(a) Parallel resistors – general case; (b)* R_E *is the equivalent single resistor of* R_1, R_2, R_3, R_4, . . . , R_N *in parallels*

$$I_S = I_1 + I_2 + I_2 + I_4 + \dots + I_N$$

Example 3.7   *For the circuit shown in Figure 3.10(a), calculate:*

(a)   *the current flowing through* R_1
(b)   *the current flowing through* R_2
(c)   *the supply current* I_S
(d)   *the equivalent single series resistor.*

(a)   *Current through* R_1
The voltage across R_1 is V_S, therefore:

$$I_1 = \frac{V_S}{R_1} = \frac{12\,V}{10\,\Omega} = 1.2\,A$$

(b)   *Current through* R_2
The voltage across R_2 also V_S, therefore:

$$I_2 = \frac{V_S}{R_2} = \frac{12\,V}{20\,\Omega} = 0.6\,A$$

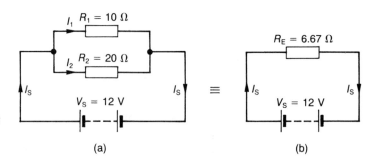

Figure 3.10 *Circuits for Example 3.7. (b) is the electrical equivalent of (a)*

(c)   Supply current I_S

$$I_S = I_1 + I_2 = 1.2\,\text{A} + 0.6\,\text{A} = 1.8\,\text{A}$$

(d)   Equivalent single resistor R_E

This is shown in Figure 3.10(b) and is calculated using R_1 and R_2 in the product-over-sum formula:

$$R_E = \frac{10 \times 20}{10 + 20} = \frac{200}{30} = 6.67\,\Omega$$

We can use this value of R_E, together with V_S, to recalculate I_S and check the correctness of the value calculated in (c) above:

$$I_S = \frac{V_S}{R_E} = \frac{12\,\text{V}}{6.67\,\Omega} = 1.8\,\text{A}$$

This confirms the value obtained in (c).

Example 3.8   *Resistors of value 1 Ω, 2 Ω, 3 Ω and 4 Ω are connected all in parallel across a 6 V battery. Calculate:*

(a)   the current drawn from the battery
(b)   the power dissipated by each resistor.

The circuit is drawn in Figure 3.11.

(a)   To calculate I_s

The first step is to calculate the equivalent single resistor as seen by the battery:

$$\frac{1}{R_E} = \frac{1}{R_1} + \frac{1}{R_2} + \frac{1}{R_3} + \frac{1}{R_4}$$
$$= \frac{1}{1} + \frac{1}{2} + \frac{1}{3} + \frac{1}{4}$$
$$= 1.0 + 0.5 + 0.333 + 0.25$$
$$= 2.083$$

Therefore, $R_E = 1/2.083 = 0.48\,\Omega$.
Using Ohm's law:

$$I_S = \frac{V_S}{R_E} = \frac{6\,\text{V}}{0.48\,\Omega} = 12.5\,\text{A}$$

(b)   Resistor power dissipation

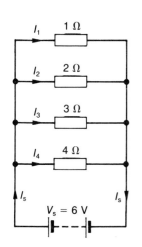

Figure 3.11 *Circuit for Example 3.8*

$1\,\Omega$ $\text{Power} = \dfrac{V_S^2}{R_{1\,\Omega}} = \dfrac{6^2}{1} = 36\,\text{W}$

Similarly, for the other three resistors:

2 Ω   Power = $6^2/2 = 18$ W
3 Ω   Power = $6^2/3 = 12$ W
4 Ω   Power = $6^2/4 = \ \ 9$ W

Total power dissipated = 75 W.
   The total power could also be calculated from:

$$\text{Total power} = \frac{V_S^2}{R_E} = \frac{36}{0.48} = 75 \text{ W}$$

3.3 Series–parallel networks

Example 3.9   *For the circuit shown in Figure 3.12, calculate the current drawn from the battery and the power dissipated in the 30 Ω resistor.*

There are several equally good ways of tackling this problem, all relying on little more than the simple application of Ohm's law. It usually makes the problem easier if the battery negative terminal is imagined as being at earth (zero) potential. Accordingly, Figure 3.12 has the battery negative terminal connected to an earth symbol and, further, the sections of the circuit operating at different potentials are indicated appropriately by solid, broken or dotted lines. We shall be particularly interested in the voltages V_A and V_B because we can use these to find the power in the 30 Ω resistor.
   We shall solve this problem by taking a series of logical steps as follows:

1 Find the equivalent single resistor, R_E, to represent the whole circuit.
2 With V_S and R_E both known it is a simple Ohm's law manipulation to find I_s.

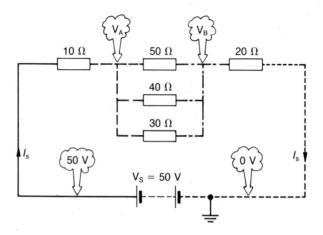

Figure 3.12 *Series and parallel resistors*

3 With I_S flowing through all of the resistors, the voltages V_A and V_B can be readily deduced.

Step 1   Find R_E
This first entails changing the three parallel resistors of $50\,\Omega$, $40\,\Omega$ and $30\,\Omega$ into an equivalent single series resistor, R_F:

$$\frac{1}{R_P} = \frac{1}{50} + \frac{1}{40} + \frac{1}{30}$$
$$= 0.02 + 0.025 + 0.0333$$
$$= 0.0783$$

Therefore:

$$R_P = \frac{1}{0.0783} = 12.7\,\Omega$$

The circuit of Figure 3.12 can now be redrawn as Figure 3.13.

$$R_E = (10 + 12.77 + 20)\ \Omega = 42.77\ \Omega$$

Step 2   Find I_S

$$I_S = \frac{V_S}{R_E} = \frac{50\,\text{V}}{42.77\,\Omega} = 1.169\,\text{A}$$

Step 3   Find V_A and V_B and hence the power in the $30\,\Omega$ resistor

$$V_A = V_B - \text{volt drop across } 10\,\Omega$$
$$= V_S - I_S \times 10\,\Omega$$
$$= 50\,\text{V} - 1.169\,\text{A} \times 10\,\Omega$$
$$= 50\,\text{V} - 11.69\,\text{V}$$
$$= 38.31\,\text{V}$$

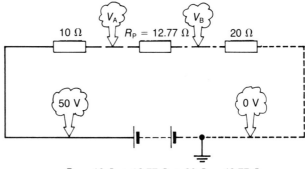

Figure 3.13 *Equivalent circuit of Figure 3.12*

$$R_E = 10\ \Omega + 12.77\ \Omega + 20\ \Omega = 42.77\ \Omega$$

$V_B = I_S \times 20\,\Omega$ above earth potential

$$= 1.169\,\text{A} \times 20\,\Omega$$

$$= 23.38\,\text{V}$$

The power dissipated by the $30\,\Omega$ resistor is given by

(voltage across $30\,\Omega$ resistor)$^2/30\,\Omega$ watts

$$P_{30\,\Omega} = \frac{(V_A - V_B)^2}{30\,\Omega}$$

$$= \frac{(38.31 - 23.38)^2}{30}$$

$$= \frac{(14.95)^2}{30}$$

$$= 7.43\,\text{W}$$

3.4 Kirchhoff's laws

For the circuits we have considered thus far, Ohm's law has been a sufficiently powerful tool to enable us to determine the current flowing and the potentials at any point. The reason for this is that the circuits have comprised only a single battery and resistors forming a single loop. Should we have a circuit of more than one loop, such as Figure 3.14, we find the two laws attributed to Kirchhoff to be more appropriate.

Figure 3.14 has three loops: ABEF, BCDE and ACDF. We start the analysis of the circuit by marking on the circuit diagram the directions of the currents I_1, I_2 and I_3. It does not matter if we guess wrongly; the answer will simply turn out to be negative telling us that the current flows the other way.

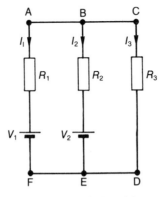

The directions of currents I_1, I_2 and I_3 are initially best estimates. If the current's flow is in the opposite direction their calculated values will simply be negative.

Figure 3.14 *Kirchhoff's laws*

Kirchhoff's First Law effectively states that the amount of current flowing into a circuit junction must be equal to the amount of current flowing away from the junction. For example, if we apply this law to junction B in Figure 3.14, we have:

$$I_1 + I_2 \qquad = \qquad I_3 \qquad\qquad [3.7]$$

Current into   =   Current away
junction B   from junction B

Kirchhoff's Second Law effectively states that the current flowing around a closed loop causes volt drops across the various components in the loop. The sum of these volt drops must equal the applied voltage causing the loop current to flow. For example, applying this law to the left and right loops of Figure 3.14 we can produced two voltage equations. But, in producing these two equations, one for each loop, we must be very careful to follow a convention we set for determining the sign of the various currents and voltages.

Let us apply the second law to the loop BCDE, which is reproduced, for convenience, as Figure 3.15. The basic equations can be formed using the following criterion:

The net voltage source in any loop = the sum of the voltage drops
                                              around that loop

The only voltage source in loop BCDE is V_2, which is in a direction which causes a conventional current to flow from V_2 in a positive (clockwise) direction.

The volt drops across the resistors are $+(I_2 \times R_2)$ for the clockwise current, I_2, flowing through resistor R_2 and $+(I_3 \times R_3)$ for the clockwise current, I_3, flowing through R_3. Note that both of these volt drops are regarded as being positive.

So we can now form the voltage equation:

$$+V_2 = +I_2R_2 + I_3R_3 \qquad\qquad [3.8]$$

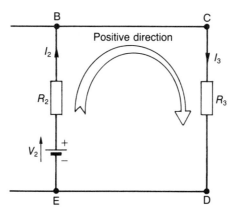

Figure 3.15 *The arrow is drawn to represent the positive direction of current flow and battery e.m.f. in loop BCDE of Figure 3.14*

Applying the second law to the loop ABEF enables us to produce a second voltage equation. First we redraw the loop as Figure 3.16 and add the positive direction arrow. Once again:

Resultant loop source volts = sum of volt drops around loop

V_1 and V_2 are both acting in this loop, but in opposition to each other. V_1 is acting in the positive direction but V_2 is acting in the negative (anticlockwise) direction. V_2 therefore is given a negative sign, and the resultant loop voltage to form the left-hand side of our voltage equation is:

$$+V_1 - V_2$$

The volt drops around the loop are $+(I_1R_1)$ and $-(I_2R_2)$. Notice how I_2 opposes our clockwise positive convention and therefore makes the volt drop across R_2 negative. The right-hand side of our loop voltage equation becomes:

$$+I_1R_1 - I_2R_2$$

The complete loop voltage equation is written as:

$$V_1 - V_2 = I_1R_1 - I_1R_2 \qquad [3.9]$$

Since we usually know the value of the circuit components, we are left with only the three curents I_1, I_2 and I_3 as the unknown quantities. But, the three equations (3.7), (3.8) and (3.9) can be solved simultaneously to determine the values of the three currents. Should a current have a negative value, it simply means that it is flowing in the opposite direction to that expected.

Example 3.10   *For the circuit shown in Figure 3.17, calculate the currents flowing in each of the three resistors.*

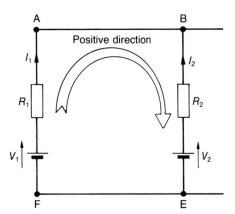

Figure 3.16 *Loop ABEF of Figure 3.14 is also marked to show the positive direction convention for currents and e.m.f.s*

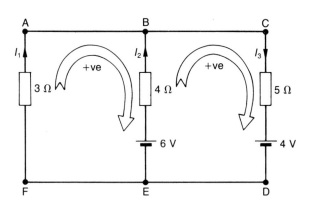

Figure 3.17 *Circuit for Example 3.10*

The most likely directions of current flowing are guessed and marked on the circuit together with the positive direction arrows in each loop. Note that a positive clockwise convention has been used for both loops, although this is not mandatory.

By Kirchhoff's first law:

$$I_1 + I_2 = I_3 \qquad [3.10]$$

By Kirchhoff's second law in the left-hand loop:

$$-6\,\text{V} = +I_1 \times 3\,\Omega - I_2 \times 4\,\Omega$$
$$-6\,\text{V} = 3I_1 - 4I_2 \qquad [3.11]$$

By Kirchhoff's second law in the right-hand loop:

$$+6\,\text{V} + 4\,\text{V} = +I_2 \times 4\,\Omega + I_3 \times 5\,\Omega$$
$$10\,\text{V} = 4I_2 + 5I_3 \qquad [3.12]$$

We now have to solve these three equations for I_1, I_2 and I_3.

Step 1   Substitute [3.10] in [3.12] and get:

$$10 = 4I_2 + 5(I_1 + I_2)$$
$$10 = 4I_2 + 5I_1 + 5I_2$$
$$10 = 5I_1 + 9I_2$$

or

$$I_1 = 10/5 - 9I_2$$
$$I_1 = 2 - 1.8I_2 \qquad [3.13]$$

Step 2   Substitute [3.13] in [3.11] and get:

$$-6\,\text{V} = 3(2 - 1.8I_2) - 4I_2$$
$$-6\,\text{V} = 6 - 5.4I_2 - 4I_2$$

and

$$I_2 = -12/-9.4 = 1.277\,\text{A}$$

Step 3   Substitute $I_2 = 1.277$ in [3.13] and get:

$$I_1 = 2 - 1.8 \times 1.277$$
$$I_1 = -0.299\,\text{A}$$

Step 4   From [3.10]:

$$I_3 = -0.299 + 1.277$$
$$I_3 = 0.978\,\text{A}$$

Summary:

$I_1 = 0.299$ A flowing from A to F
$I_2 = 1.277$ A flowing from E to B
$I_3 = 0.978$ A flowing from C to D

Example 3.11   *For Figure 3.18, determine the currents flowing through each of the resistors.*

Once again we decide for each of the loops ABEF and BCDE that a clockwise direction will indicate a positive current flow. We then apply the two Kirchhoff laws to produce three simultaneous equations which we can solve for I_1, I_2 and I_3.

Applying Kirchhoff's first law to junction E:

$$I_3 = I_1 + I_2 \tag{3.14}$$

Applying Kirchhoff's second law to loop ABEF:

$$-6\,\text{V} = -50I_2 - 100I_3$$

or

$$-3\,\text{V} = -25I_2 - 50I_3 \tag{3.15}$$

Applying Kirchhoff's second law to loop BCDE:

$$+8\,\text{V} = 40I_1 + 100I_3$$

or

$$2\,\text{V} = 10I_1 + 25I_3 \tag{3.16}$$

Solving [3.14], [3.15] and [3.16] simultaneously for the three currents we obtain:

$I_1 = 0.545$ A or 54.5 mA
$I_2 = 0.0037$ A or   3.7 mA
$I_3 = 0.0582$ A or 58.2 mA

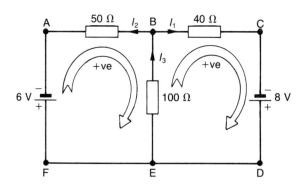

Figure 3.18 *Circuit for Example 3.11*

3.5 Thevenin's theorem

This is another tool which can be used not only for the analysis of d.c. circuits but also in other applications where it is convenient to replace a complicated circuit by a simple constant voltage source together with a series resistor.

Suppose that we encounter a complicated d.c. network and that we need to know the current flowing through only one particular resistor. Figure 3.19 shows this situation. R is the single resistor of interest and I is the current it is passing. We imagine that R is connected into its parent circuit through terminals marked A and B. If we now remove the resistor R, and measure the voltage that exists between A and B, we call this the *open circuit voltage*, V_{OC}. This is shown by Figure 3.20(a). We can undertake a further measurement, between A and B, called the *input resistance*, R_{in}. For this measurement, R is removed, as shown in Figure 3.20(b). The circuit in Figure 3.19 can be replaced by Figure 3.21.

Knowing the value of V_{OC}, R_{in} and R, it is a relatively easy matter of applying Ohm's law to determine the current I flowing through R:

$$I = \frac{\text{circuit e.m.f.}}{\text{circuit resistance}}$$

$$I = \frac{V_{OC}}{R + R_{in}} \text{ amps} \qquad [3.17]$$

A more formal statement of Thevenin's theorem would be:

The current in a load resistance connected to two terminals, A and B, of a network of resistances and generators is the same as if this load

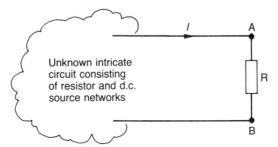

Figure 3.19 *A circuit to illustrate Thevenin's theorem*

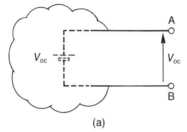
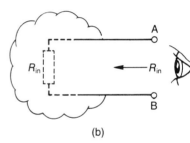

Figure 3.20 *(a) Open circuit voltage; (b) input resistance 'looking in' between open circuited terminals A and B*

(a)   (b)

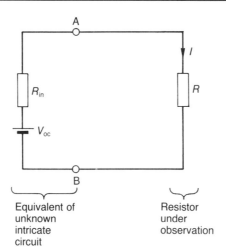

Figure 3.21 *Thevenin equivalent circuit of Figure 3.19*

resistance were connected to a simple constant voltage generator, whose e.m.f. is the open circuit voltage measured across A and B, and whose internal resistance is the resistance of the network looking back into the terminals A and B with all generators replaced by resistances equal to their internal resistances.

Example 3.12   *Use Thevenin's theorem to find the current flowing through the 10 Ω resistor in the circuit shown in Figure 3.22.*

Step 1   Remove the 10 Ω resistor leaving the terminals A and B open circuited. Now calculate the voltage, V_{OC}, between A and B. Figure 3.23 is the resulting circuit which shows V_{OC} to be the voltage being developed across the 4 Ω resistor caused by the current, I_1, flowing around the closed loop:

$$I_1 = 10\,\text{V}/(3+4)\,\Omega = 10\,\text{V}/7\,\Omega = 1.428\,\text{A}$$

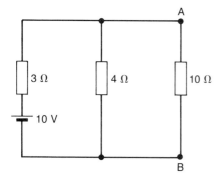

Figure 3.22

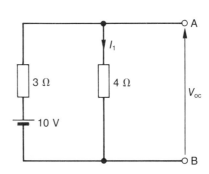

Figure 3.23

Therefore:

$$V_{OC} = 1.428\,\text{A} \times 4\,\Omega = 5.71\,\text{V}$$

Alternatively:

$$V_{OC} = 10\,\text{V} \times 4\,\Omega/(3+4)\,\Omega = 5.71\,\text{V}$$

Step 2 Now find R_{in} by redrawing Figure 3.23 as Figure 3.24 and replacing the 10 V battery with its internal resistance, which in this case is zero.

R_{in} is simply the effect of the 4 Ω resistor in parallel with the 3 Ω resistor.

Step 3 Now draw the Thevenin equivalent circuit, at the same time replacing the 10 Ω resistor between the terminals A and B. Figure 3.25 is the result.

Step 4 Calculate the current flowing through the 10 Ω resistor:

$$\begin{aligned} I &= \text{(circuit voltage)}/\text{(circuit resistance)} \\ &= 5.71\,\text{V}/(10+1.71)\,\Omega \\ &= 0.448\,\text{A or } 48.8\,\text{mA} \end{aligned}$$

Not all circuits have the same layout as the one used in Example 3.12, and the following notes are intended to help the reader in tackling the problem of determining V_{OC} and R_{in}.

3.5.1 *General determination of V_{OC} and R_{in}*

Figure 3.26(a) is the simplest case. After the removal of the load resistor from terminals A and B, the open circuit voltage is the e.m.f. of the battery. R, which includes the internal resistance of the battery, constitutes the input resistance, R_{in}.

To obtain V_{OC} for Figure 3.26(b), the circuit is best redrawn as shown by Figure 3.26(c). The three resistors form a potential divider across the 12 V

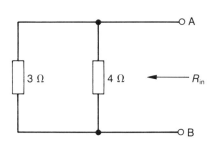

Figure 3.24

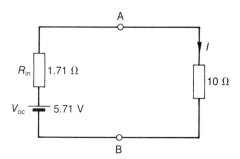

Figure 3.25

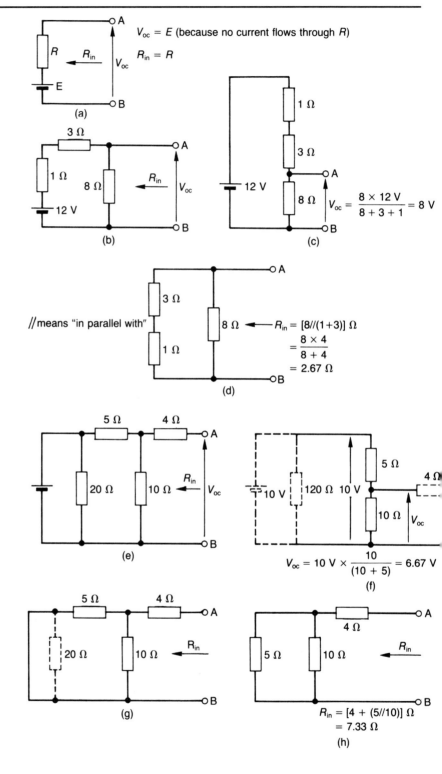

Figure 3.26 *Determination of* V_{OC} *and* R_{in}

battery with $V_{OC} = 8$ V appearing across the 8 Ω resistor. Figure 3.26(b) is redrawn as Figure 3.26(d), from which it is apparent that R_{in} looks like the 8 Ω resistor in parallel with the other two giving a value of 2.67 Ω.

The circuit in Figure 3.26(e) is redrawn as Figure 3.26(f). This is then used to obtain V_{OC}, noting that the 20 Ω and the 4 Ω resistors (shown dotted) have no current flowing through them and therefore do not affect V_{OC}, which is 6.67 V. To obtain R_{in}, Figure 3.26(e) is redrawn in two stages, Figure 3.26(g) and then Figure 3.26(h). It will be noted that the battery is effectively a short-circuit across the 20 Ω resistor.

Example 3.13 *Find the current flowing through the 15 Ω resistor in Figure 3.27(a).*

Step 1 Remove the 15 Ω resistor and draw Figure 3.27(b) to determine $V_{OC} = 10$ V.

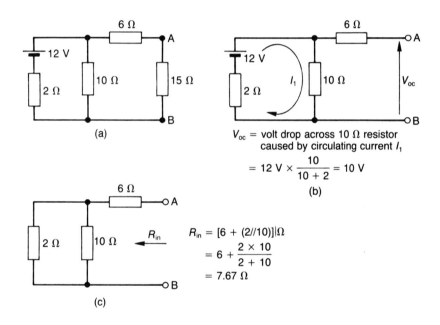

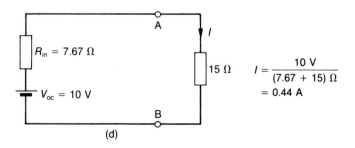

Figure 3.27 *Circuits for Example 3.13*

Step 2 Draw Figure 3.27(c) and determine $R_{in} = 7.67\,\Omega$.

Step 3 Draw the Thevenin equivalent circuit in Figure 3.27(d) with the $15\,\Omega$ resistor reconnected and determine the current it passes as $0.44\,A$.

Example 3.14 *Find the current flowing in the $5\,\Omega$ resistor in the circuit shown in Figure 3.28(a).*

Step 1 Remove the $5\,\Omega$ resistor and redraw the circuit as Figure 3.28(b) in order to calculate V_{OC} across terminals A and B:

$$V_{OC} = 6\,V - \text{p.d. across the } R_{2\Omega} \text{ resistor}$$

(Alternatively, $V_{OC} = 3\,V +$ p.d. across the $R_{4\Omega}$ resistor.)
    We now need to find I_1. Using Kirchhoff's second law in the left loop:

$$6\,V - 3\,V = 2I_1 + 4I_1$$

Therefore:

$$I_1 = 3/6 = 0.5\,A$$

We can use this value of I_1 to determine V_{OC}:

$$V_{OC} = 6\,V - (2\,\Omega \times 0.5\,A) = 5\,V$$

(Alternatively, $V_{OC} = 3\,V + (4\,\Omega \times 0.5\,A) = 5\,V$.)

Step 2 Replace the batteries by their internal resistances (both of them zero in this case), and hence draw Figure 3.28(c) which is used to determine R_{in}:

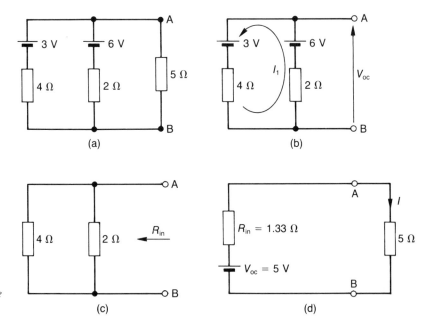

Figure 3.28 *Circuits for Example 3.14*

$R_{in} = 2\,\Omega$ in parallel with $4\,\Omega = 1.33\,\Omega$

Step 3 Draw the Thevenin equivalent circuit as in Figure 3.28(d), which includes the reinserted $5\,\Omega$ resistor, and hence calculate the current I:

$$I = 5\,\text{V}/(1.33 + 5)\,\Omega = 0.79\,\text{A}$$

Exercises

3.1 For the circuit shown in Figure 3.29, calculate:
    (a) the total circuit resistance
    (b) the circuit current
    (c) the total power dissipated.

3.2 For the circuit shown in Figure 3.30, calculate:
    (a) the voltage at X
    (b) the voltage at Y
    (c) the power dissipated by the $50\,\Omega$ resistor.

3.3 For the circuit shown in Figure 3.31, calculate:
    (a) the total circuit resistance
    (b) the current drawn from the battery
    (c) the power dissipated by the $20\,\Omega$ resistor.

3.4 For the circuit shown in Figure 3.32, calculate:
    (a) the circuit current, I_S
    (b) the voltage potentials at A and B
    (c) the current flowing through the $10\,\Omega$ resistor
    (d) the power dissipated by the $20\,\Omega$ resistor.

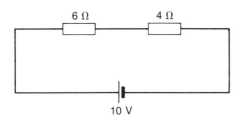

Figure 3.29

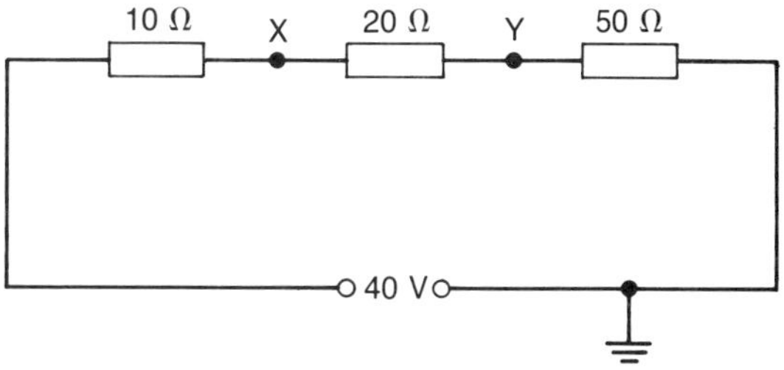

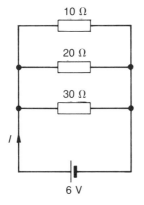

Figure 3.31

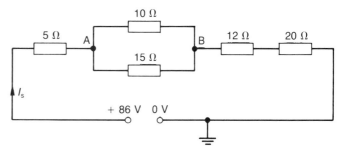

Figure 3.30

Figure 3.32

3.5   Figure 3.33 shows how three identical lamps, A, B and C, are lit by a current, I, being drawn from a constant voltage supply of V volts.
   (a)   If a piece of conducting wire is used to short point C to point D, state the effect on the brightness of lamps L1, L2 and L3.
   (b)   Suppose that the short in (a) is removed and SW1 then closed to include the extra resistor, R, in the circuit. Will L3 be more or less bright with the closing of SW1? And will the circuit current have increased or decreased?
   (c)   With point F shorted to point G, will the lamps L1 and L2 be more bright, less bright or of unchanged brightness?
3.6   For the circuit shown in Figure 3.34, use Kirchhoff's laws to calculate the current passed by the 5 Ω resistor.
3.7   For the circuit shown in Figure 3.35, use Kirchhoff's laws to calculate:
   (a)   the current flowing through each battery
   (b)   the voltage across the 10 Ω resistor
   (c)   the power dissipated in the 10 Ω resistor.
3.8   For the circuit in Figure 3.36, use Kirchhoff's laws to determine the power dissipated by the 1 Ω resistor.
3.9   If in Figure 3.36 the 8 V battery were reversed, calculate the new power dissipated by the 1 Ω resistor.

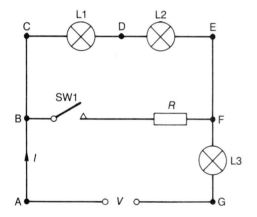

Figure 3.33

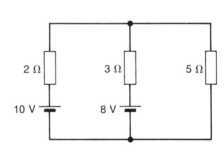

Figure 3.34

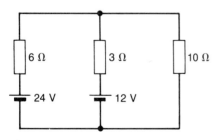

Figure 3.35

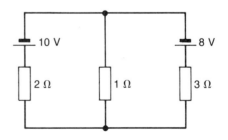

Figure 3.36

3.10 Use Thevenin's theorem to find the current flowing in the 5 Ω resistor of Figure 3.34 of question 3.6.

3.11 Assuming that the voltage source has negligible internal resistance, use Thevenin's theorem to find the current flowing in the 4 Ω resistor of Figure 3.37.

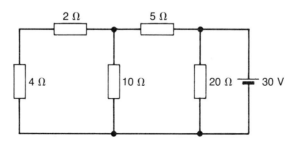

Figure 3.37

4 Alternating current theory

4.1 Introduction

In Section 2.3, the generation of continually reversing or alternating currents (a.c.) was discussed. The a.c. generator is called an *alternator*. The usual method of generating this form of electrical energy is by mechanically rotating a loop conductor in a magnetic field and making use of the current that is induced into the loop. The magnitude versus time relationship for the generated alternating current is called its *waveform*. The waveform usually associated with a current generated by a rotating loop is a *sine wave* or *sinusoidal waveform*.

4.2 The sinusoidal waveform

A waveform is a graphical record or representation of how a voltage or current amplitude varies with time. Because of the way electricity is generated and distributed, a very common waveform encountered in electrical and electronic engineering is the *sine wave*.

Figure 4.1 shows a rotating loop situated within a vertical permanent magnetic field flux. The details of how a current is induced in and extracted from the loop is covered in Chapter 2. The purpose of Figure 4.1 is to show how part of the rotating loop, O to A, is analogous to the rotating voltage vector, OA, in Figure 4.2(a). In Figure 4.1, the loop is shown in the position where θ is $0°$, where the flux cutting is zero and the loop-induced current and voltage are also zero. An anticlockwise rotation to $\theta = 90°$ is the position of maximum rate of cutting flux and maximum induced loop voltage and current. In Figure 4.2, the vector, OA, of length V_{max}, is rotating at a constant angular velocity, ω radians/second. It is generating an instantaneous vertical projection, v, where $v = V_{max} \sin \theta$. This is also the expression for the instantaneous voltage generated by the rotating loop of Figure 4.1.

Figure 4.2(b) is derived from Figure 4.1(a). It is a plot of v versus θ and is a sine wave. Alternatively, since $\theta = \omega t$, where ω is constant, the horizontal axis can be labelled in units of time.

Figure 4.3 helps to explain some of the terms used with sinusoidal and other waveforms. Note that the waveform is drawn symmetrically above and below the zero voltage line and forms the so-called positive and negative half-cycles. In practice, during the positive half-cycle, current

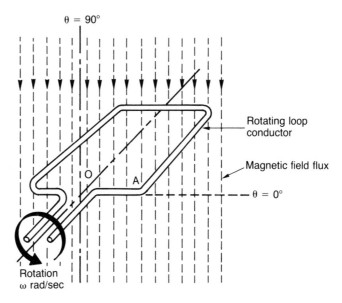

Figure 4.1 *Rotating a loop conductor in a magnetic field generates an instantaneous e.m.f. proportional to the sine of the angle, θ, at which the loop is cutting the flux (see also Section 2.3)*

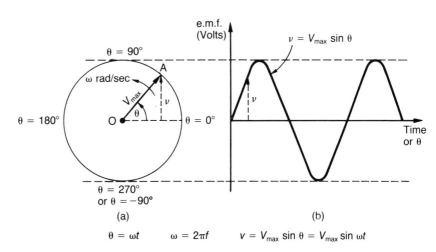

Figure 4.2 *The generation of a sine wave by a vector of length V_{max} rotating at a constant angular velocity ω*

flows one way around a circuit but is reversed during the negative half-cycle. During a single cycle the current is said to be *bidirectional*.

Not all alternating bidirectional waveforms are sinusoidal and Figure 4.4 shows a selection of other important waveforms encountered in electronic engineering.

Example 4.1 *An alternating bidirectional voltage has a frequency of 50 Hz and an amplitude of 340 V. Calculate the periodic time of the waveform and its peak-to-peak value.*

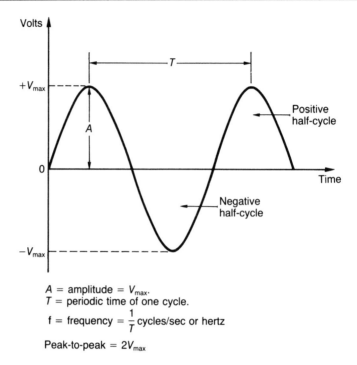

A = amplitude = V_{max}.
T = periodic time of one cycle.
f = frequency = $\dfrac{1}{T}$ cycles/sec or hertz

Peak-to-peak = $2V_{max}$

Figure 4.3 *Terms used with a bidirectional voltage waveform*

The periodic time is the reciprocal of the frequency:

$$T = \frac{1}{f} \qquad T = \frac{1}{50\,\text{Hz}} = 0.02\,\text{s}$$

The peak-to-peak value is simply twice the amplitude at 680 V.

4.2.1 *The average value of a half-cycle of a sine wave*

The average value of a complete cycle of a sine wave with its positive and negative half-cycle both present must be zero – the two half-cycles cancel each other. However, there are some electrical applications that are not sensitive to which way the current is flowing so the positive and negative half-cycles do not cancel. For example, some instruments respond to the average current flowing through them irrespective of its direction. For this reason it is important that we can estimate the average value of alternating currents which have different waveforms. We shall confine our studies to the sine wave.

Figure 4.5 shows the waveform of a voltage given by the mathematical equation $v = V_{max} \sin \theta$, where v is the instantaneous value of the voltage at an instantaneous angle, θ. V_{max} is the amplitude of the waveform. Because of the symmetry of a sine wave, we need only work out the average value for a half-cycle; this will be the same as for the whole cycle.

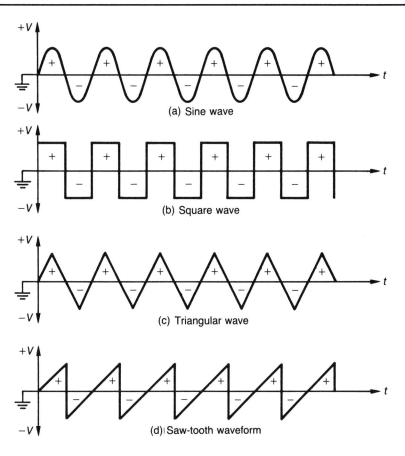

Figure 4.4 *Examples of bidirectional waveforms, all of the same amplitude and frequency*

(a) Sine wave

(b) Square wave

(c) Triangular wave

(d) Saw-tooth waveform

The average value of the half-cycle is given by the height of the rectangle OABC, where the area of the rectangle is equal to the area of the positive half-cycle.

Thus, we can write:

$$\pi \times V_{av} = \int_0^\pi v \, d\theta$$

$$= \int_0^\pi V_{max} \sin \theta \, d\theta$$

$$= V_{max}[\cos \theta]_0^\pi$$

$$= V_{max}[-\cos \pi + \cos 0]$$

$$= V_{max}[-(-1) + 1]$$

$$= 2V_{max}$$

Therefore:

$$V_{av} = (2/\pi) V_{max} = 0.637 V_{max} \qquad [4.1]$$

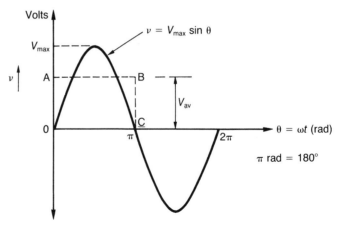

The average value of a half-cycle is given by
V_{av} where the area of rectangle OABC is equal
to the area under the sine curve 0 to C.

Figure 4.5 *The average value of a sine wave*

$V_{av} = 0.637 V_{max}$

4.2.2 The r.m.s. value of a sinusoidal voltage or current (V_{rms} or I_{rms})

The letters r.m.s. stand for the phrase *root mean square*. In engineering terms, the r.m.s. value of a current or voltage means that it will produce the same power or heating effect as a d.c. current or voltage of the same nominal value. For example, an a.c. voltage of 10 V r.m.s. will produce the same power as a battery voltage of 10 V d.c. The power developed by a current or voltage is proportional to their value squared. The method of determining an r.m.s. value is shown by Figure 4.6, where:

$$V_{rms} = \sqrt{(v^2)_{av}} \qquad [4.2]$$

Now:

$$v^2 = V_{max}^2 \sin^2 \theta$$
$$= 0.5 V_{max}^2 (1 - \cos 2\theta)$$

The area under the v^2 curve is given by:

$$\int_0^\pi 0.5 V_{max}^2 (1 - \cos 2\theta) \, d\theta$$
$$= 0.5 V_{max}^2 [\theta - 0.5 \sin 2\theta]_0^\pi$$
$$= 0.5 V_{max}^2 [\pi + 0.5 \sin 2\pi - 0 + 0.5 \sin 0]$$
$$- 0.5 V_{max}^2 [\pi + 0 - 0 + 0]$$
$$= (\pi/2) V_{max}^2$$

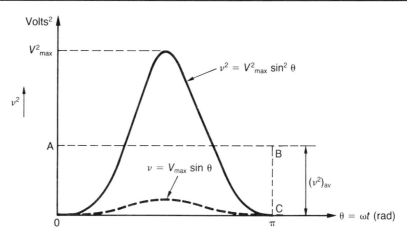

Volts²

V^2_{max}

$v^2 = V^2_{max} \sin^2 \theta$

v^2

A

B

$v = V_{max} \sin \theta$

$(v^2)_{av}$

C

$\theta = \omega t$ (rad)

0   π

Figure 4.6 *The r.m.s. value of an a.c. voltage (or current) has the same heating effect as a d.c. voltage (or current) of the same nominal value*

The power or heat produced is proportional to (voltage)². The heat produced by a sinusoidal voltage, $v = V_{max} \sin \theta$, is proportional to the average value of $v^2 = (v^2)_{av}$.

$\sqrt{(v^2)_{av}}$ is called the "root-mean-square" value.

$$V_{rms} = \sqrt{(v^2)_{av}} = 0.707 \, V_{max}$$

Therefore:

$$(v^2)_{av} \times \pi = 0.5\pi V^2_{max}$$
$$(v^2)_{av} = 0.5 V^2_{max}$$

Substituting this in Equation 4.2 we have:

$$V_{rms} = \sqrt{(v^2)_{av}} = \sqrt{(0.5 V^2_{max})} = 0.707 V_{max} \qquad [4.3]$$

Example 4.2   *The mains voltage has sinusoidal waveform and is quoted as having a value of 230 V. Calculate:*

(a)   the peak value of the waveform
(b)   the half-cycle average value
(c)   the power it would dissipate in a 500 Ω resistor.

Note: In the absence of other information, the value given for an a.c. voltage is always assumed to be the r.m.s. value.

(a)   The peak value, or amplitude, using Equation 4.3 would be:

$230/0.707 = 325.3$ V

(b)   The average value, using Equation 4.1, would be:

$325.3 \times 0.637 = 207.2 \, \text{V}$

(c)   The power dissipated in a $500 \, \Omega$ resistor would be:

$(V_{rms}^2)/R = 230^2/500 = 105.8 \, \text{W}$

4.3 Reactance (X)

Equation 1.10 in Chapter 1 is repeated here for convenience:

$$e = -L \frac{\mathrm{d}i}{\mathrm{d}t} \quad \text{when } \delta t \text{ is small.} \tag{4.4}$$

This is a derivation from Faraday's laws and says that the voltage across an inductance, L henries, is given by L multiplied by the rate the current flowing through the inductance is changing.

Figure 4.7(a) shows the waveform of a sinusoidally varying current having the equation:

$$i = I_{max} \sin \omega t \tag{4.5}$$

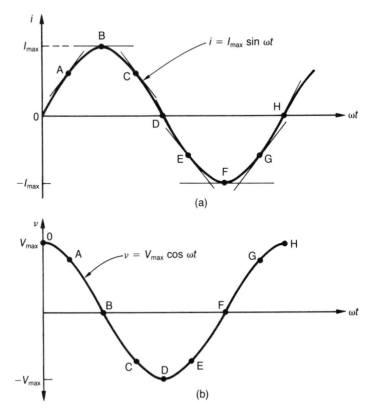

Figure 4.7 *The current flowing through a pure inductance lags the voltage across it by 90°. (a) Current sine wave with tangents drawn at points A to H; (b) voltage cosine wave drawn by plotting the slopes of the tangents in (a)*

where i is the instantaneous current flowing at any instant t.

If tangents are drawn to each of the points A to H on this curve and their slopes plotted, the cosine curve of Figure 4.7(b) is obtained. It will be noted that the cosine wave of (b) is no more than the sine wave of (a) advanced by 90°. Also, since the cosine wave (b) has been derived from the slope of (a), in other words, from di/dt, the cosine wave is proportional to the *voltage* across the inductance. The equation for the instantaneous voltage is thus:

$$v = V_{max} \cos \omega t \qquad [4.6]$$

If we divide the voltage across a component by the current flowing through it, we expect to obtain the value, in ohms, of the component's resistance. But in the case of pure inductance, it is not quite so simple because the voltage and current are 90° out of phase; they are said to be in *quadrature*. The result of dividing a voltage by a quadrature current is still measured in ohms but is called *reactance* rather than resistance.

If we represented a resistance of $10\,\Omega$ by a horizontal line, or vector, 10 cm long, we could correctly represent an *inductive reactance* of $10\,\Omega$ by a vector 10 cm long but turned anticlockwise through 90° into a vertical position. (We shall be discussing the matter of the vector representation of electrical quantities in more detail later in this chapter.)

We can determine a relationship between the reactance of a coil in ohms and its inductance in henries as follows.

We know that the equation for a sinusoidal current is:

$$i = I_{max} \sin \omega t$$

and if we differentiate this with respect to time we obtain the slope of the tangent at any point to the curve, in other words, di/dt:

$$\frac{di}{dt} = \omega I_{max} \cos \omega t$$

We can substitute this value of di/dt in Equation 4.4 and, ignoring the minus sign, we can write the magnitude of e as:

$$e = \frac{L di}{dt} = \omega L I_{max} \cos \omega t$$

And since $\cos \theta = \sin(\theta + 90°)$:

$$e = \omega L I_{max} \sin(\omega t + 90°)$$

But, from Equation 4.5, $I_{max} \sin(\omega t + 90°)$ is the same as i, but advanced by 90°, so we can write:

$$e = \omega L i \text{ in magnitude}$$

Now, reactance is simply e divided by i so we can say:

$$\text{Inductive reactance}, X_L = \frac{e}{i} = \omega L \text{ ohms} \qquad [4.7]$$

The above argument has been limited to the inductive reactance of, say, a coil. However, the situation is similar for the *capacitive reactance* of a capacitor. Whereas the inductance has a current flowing through it which *lags by 90°* the voltage applied across it, the situation with capacitor is just the opposite. The current flowing into a capacitor *leads by 90°* the voltage applied across it.

The equation for the instantaneous flow of current into a capacitor is given by:

$$i = C\frac{dv}{dt} \tag{4.8}$$

where C is the capacitance in farads and dv/dt is the rate of change of voltage between the plates of the capacitor.

By a similar treatment to that for the inductance it can be shown that:

$$\text{Capacitive reactance, } X_c = \frac{1}{\omega C} \text{ ohms} \tag{4.9}$$

Equations 4.7 and 4.9 involve the use of ω, the angular frequency of the voltage and current being supplied to the reactances. ω is measured in radians per second and it is often more convenient to express this in terms of frequency. Since there are 2π radians in each revolution or cycle, and there are f cycles per second, we can write:

$$\omega = 2\pi f \tag{4.10}$$

This means that both of the reactances, $X_L = 2\pi f L$ and $X_c = 1/(2\pi f C)$, are frequency sensitive.

Example 4.3 *Calculate the current flowing through an inductance of 15.9 mH connected to a 10 V, 50 Hz supply.*

Inductive reactance, X_L, is given by:

$$X_L = 2\pi f L$$
$$= 2\pi \times 50 \times 15.9 \times 10^{-3}$$
$$= 5\,\Omega$$

Current flowing = volts/ohms = $10\,\text{V}/5\,\Omega = 2\,\text{A}$.

Example 4.4 *Calculate the current flowing into a capacitor of 339 pF connected to a 470 kHz, 25 V supply.*

Capacitive reactance, X_C, is given by:

$$X_c = \frac{1}{2\pi f C}$$
$$= 1/(2\pi \times 470 \times 10^{-3} \times 339 \times 10^{-12})$$
$$= 1000\,\Omega.$$

Current flowing = volts/ohms = $25\,\text{V}/1000\,\Omega = 25\,\text{mA}$.

We can now leave the mathematical treatment of reactance and consider its physical aspects. Figure 4.8 is the electrical diagram for a coil of

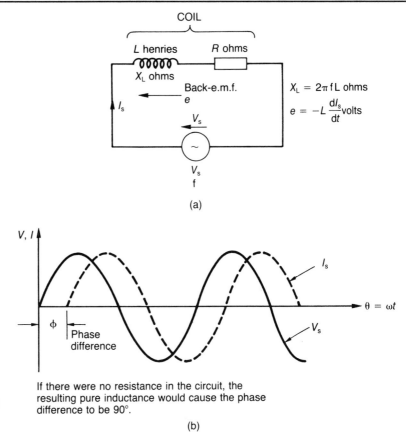

COIL

L henries   R ohms
X_L ohms

Back-e.m.f.
e

$X_L = 2\pi f L$ ohms

$e = -L \dfrac{dI_s}{dt}$ volts

I_s

V_s

V_s
f

(a)

V, I

I_s

$\theta = \omega t$

ϕ

Phase
difference

V_s

If there were no resistance in the circuit, the
resulting pure inductance would cause the phase
difference to be 90°.

(b)

Figure 4.8 *(a) Self-induced back-e.m.f. opposes* V_S*; (b) action of (a) causes the current,* I_S *to be phase delayed compared with* V_S

wire forming an inductor. Since it is impossible to wind a coil without having some resistance present in the wire, we consider all the circuit resistance together as R, and have the pure inductance, L, shown separately. The supply voltage, V_S, is attempting to force the supply current, I_S, around the circuit not only against the standard ohmic resistance, $R\,\Omega$, but also against the frequency-dependent reactance, $2\pi f L\,\Omega$, of the inductance.

The resistance, R, presents a constant value for all currents from d.c. to a.c. at high frequency. The reactance, however, is a variable quantity that depends upon the frequency, or speed of rotation, of the supplying alternator. At zero frequency, d.c. in fact, the inductive reactance, $2\pi f L$, is zero and the coil as a whole appears to be a straight piece of wire of resistance R. At high supply current frequencies, the term $2\pi f L$ can become very large to the extent that it appears as an open circuit.

The action of the inductive reactance is to oppose the continuously changing a.c. current. The coil reacts to the changing magnetic flux, caused by the changing current, by producing a self-induced back-e.m.f. which opposes the main supply voltage. It is this back-e.m.f. which prevents the immediate build-up of current in response to a voltage applied to the coil. The result is that there is a time delay, or phase difference, between V_S and I_S. This action is shown in Figure 4.8. The higher the frequency of the applied current, the greater the back-e.m.f. and the greater the value of X_L. It is the constant resistance, R, and the frequency-sensitive inductive reactance, X_L, that together form what is known as the total circuit *impedance*, Z.

4.4 Impedance (Z)

This is the combination of the resistance and the reactance of a circuit to impede the flow of an alternating current. The resistive element is constant whatever the frequency of the current alternations but the reactance varies very much with frequency. Further, it is important to appreciate that while the resistance and capacitance together form the electrical impedance of a circuit, they are not simply added arithmetically. It is because the current through a pure reactance is 90° phase displaced that the resistance and reactance must be added vectorially.

Figure 4.9 gives a hydraulic analogy which some readers may find helpful in appreciating the difference between resistance and reactance, both of which are measured in ohms.

Figure 4.9(a) shows a man walking up the sloping bottom of an empty bathing pool. In order to move forward he must exert a force to lift his weight up the slope. This lift force is the same whatever speed he proceeds up the slope. Therefore, the slope is analogous to the resistance, R, in a circuit which also has the same value for all current frequencies.

In Figure 4.9(b), the bathing pool has water in it and the man walking up the sloping bottom encounters not only a resistance to motion because of his having to lift his weight up the slope, but also a drag force which increases with speed. (Try running in a bathing pool!) The speed-variable drag force of the water is analogous to the frequency-variable inductive reactance, X_L, of the electrical circuit. Further, Figure 4.9(b) shows why it is necessary to combine the effects of the slope and the drag of the water vectorially. It is because the two are at 90° to each other. The vector diagram of the various forces is shown in Figure 4.10 and, using Pythagoras' theorem, we conclude that the impedance of a circuit, Z ohms, is given by the relationship:

$$Z^2 = R^2 + X_L^2 \qquad [4.11]$$

The reactive element, X, of the total circuit impedance, Z, need not always be inductive as provided by a coil. It can be provided by a capacitor which has a capacitive reactance, X_C. This differs from the inductive reactance in that while the latter increases with an increased frequency, X_C decreases with an increased frequency. This is evident from Equations 4.7 and 4.9: increasing f, in $X_L = 2\pi fL$, makes X_L larger but has the opposite effect on $X_C = 1/(2\pi fC)$.

Once again, we can illustrate hydraulically the effects of frequency on inductive and capacitive reactances.

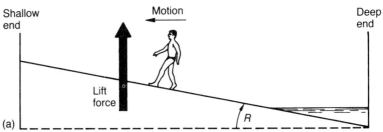

(a) Man walking from deep end to shallow end of empty pool has only the resistance caused by the sloping pool bottom, R, to overcome.

Impedance to motion is only the lift force ∝ R

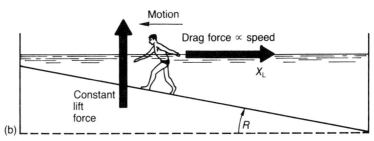

(b) With water in the pool, the man walking up the sloping bottom must not only provide a lift force to overcome the slope, he encounters an additional water drag which increases the faster he attempts to move.

Impedance to motion is the combination of the vertical lift force and the horizontal force caused by the water reaction to motion through it

Figure 4.9 *Hydraulic analogy for resistance, reactance and impedance*

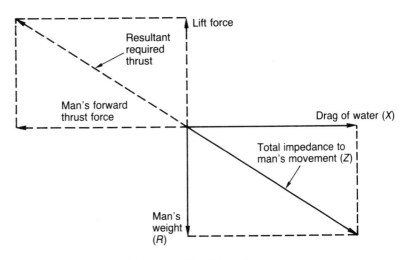

Figure 4.10 *The vectorial addition of man's weight and water drag gives the total force to be overcome to cause forward motion at a particular speed*

By Pythagoras, $Z^2 = R^2 + X^2$

Figure 4.11(a) shows the slow but stable displacement hull type of boat. The faster it tries to move through the water, the greater the retarding drag of the water on the hull. The drag is similar to inductive reactance which increases with frequency. On the other hand, Figure 4.11(b) shows a boat having a planing hull which partially lifts out of the water as the speed increases. The reduced hull contact area with the water is analogous to the reduced capacitive reactance with increased frequency.

(a) Boat with 'displacement' hull (inductance)

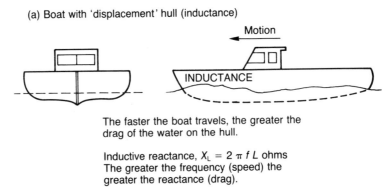

The faster the boat travels, the greater the drag of the water on the hull.

Inductive reactance, $X_L = 2 \pi f L$ ohms
The greater the frequency (speed) the greater the reactance (drag).

(b) Boat with 'planing' hull (capacitance)

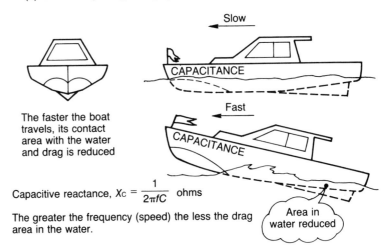

The faster the boat travels, its contact area with the water and drag is reduced

Figure 4.11 *Hydraulic analogy for the variation of inductive and capacitive reactances with frequency (speed)*

Capacitive reactance, $X_C = \dfrac{1}{2\pi f C}$ ohms

The greater the frequency (speed) the less the drag area in the water.

Area in water reduced

4.5 Phasor diagrams

Figure 4.2 illustrated how a vector of constant length, rotating at ω rad/sec, produces a vertical projection having a sinusoidal variation. Figure 4.12 extends this to a couple of vectors of different lengths, both rotating at the same angular velocity, ω, but separated by an angle ϕ. The two sinusoidal waveforms produced have the same frequency but different amplitudes. Also, there is a phase difference, ϕ, or time delay, between the two.

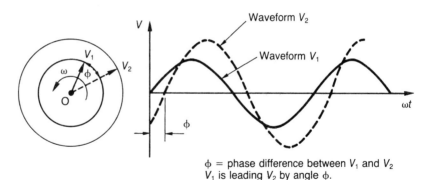

Figure 4.12 *Constant amplitude rotating vectors generate sinusoidal waveforms*

ϕ = phase difference between V_1 and V_2
V_1 is leading V_2 by angle ϕ.

If we sat on the rotating vector marked as V_2 and looked in the direction of rotation, we would see vector V_1 ahead of us, by a constant angle ϕ, but travelling at the same angular velocity, ω. Because the rotation is anticlockwise, V_1 is said to lead V_2, or V_2 lags V_1. To an observer travelling with either of the rotating vectors, the two appear to be stationary.

A set of rotating vectors, artificially held stationary, is called a *phasor diagram*. The angular velocity, ω, is taken for granted and not shown on the phasor diagram.

4.5.1 Series inductance and resistance

Figure 4.13(a) shows the circuit to be considered in this section. Figure 4.13(b) is a simplified circuit diagram where the inductive reactance and the resistance have been added vectorially to form Z, the circuit impedance. The formula used for this is given in Equation 4.11. We shall see example calculations of this type later in this chapter.

Figure 4.13(c) shows how the *voltage phasor diagram* is drawn for Figure 4.13(a). The procedure for this is as follows:

Step 1 Draw a horizontal phasor from left to right of any length to represent the circuit current, I_S. This phasor is taken as the reference direction phasor from which all angles are measured. It is logical to use the current phasor as the reference direction because the current is the one quantity which is common to all the components in the circuit: at any one time, it is flowing equally through all of them.

Step 2 Now draw the voltage phasor, V_R, which represents the volt drop across the resistor, R. Because the current flowing through a resistor and the volt drop across it must always be in-phase, that is, rise and fall together, the phasor, V_R, is drawn in-line

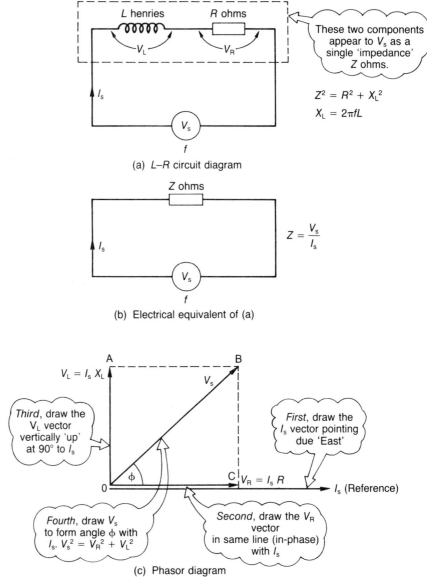

Figure 4.13 *The inductance–resistance circuit*

with I_S and in the same direction. The length of V_R can be scaled to represent its value given by $I_S \times R$.

Step 3   Draw the phasor to represent the volt drop across the inductive reactance, $V_L = I_S \times X_L$. The direction of V_L is easily decided by remembering two facts: first, the current flowing through an inductance cannot be changed immediately; and, second, the current flowing through and the voltage across a pure inductor are at 90° to each other. This means that I_S must lag the voltage, V_L. With the phasor diagram assuming an anticlockwise rotation, V_L must therefore be drawn vertically 'up' for it to be 90° ahead of I_S.

Step 4   The 'parallelogram of forces' is completed by drawing in the supply voltage, V_S, as the vectorial resultant of summing V_L and V_R.

We can apply Pythagoras' theorem to the triangle, OBC, in Figure 4.13(c), obtaining:

$$V_S^2 = V_R^2 + V_L^2 \qquad\qquad [4.12]$$

Triangle OAB is called the *voltage triangle*.

Now, if we divide volts by current, we obtain ohms; so, dividing both sides of Equation 4.12 by the common circuit current, I_S, we obtain what is called the *impedance triangle*:

$$\left(\frac{V_S}{I_S}\right)^2 = \left(\frac{V_R}{I_S}\right)^2 + \left(\frac{V_S}{I_S}\right)^2$$

or

$$Z^2 = R^2 + X_L^2 \qquad\qquad [4.13]$$

On the other hand, if we had multiplied the voltages of Equation 4.12 by I_S, we would have produced the *power triangle*:

$$(V_S I_S)^2 = (V_R I_S)^2 + (V_L I_S)^2 \qquad\qquad [4.14]$$

or

(Apparent power)^2 = (true power)^2 + (reactive power)^2

The voltage, impedance and power triangles are shown in Figure 4.14. In the impedance triangle, the phasors are all measured in ohms. In the power triangle, the true power, dissipated by the resistor, is measured in watts; the apparent power generated is measured in volt-amps, indicating that V_S and I_S are not in phase, and the 'wattless or reactive power' associated with inductance is measured in volt-amps reactive.

These triangles are all similar and are a useful aid to solving numerical problems.

Example 4.5   *Calculate the current drawn from the mains supply of 250 V, 50 Hz, when connected to a coil of resistance 40 Ω and inductance 95.5 mH.*

Figure 4.15 shows the circuit.
    Inductive reactance, X_L, is given by:

$$X_L = 2\pi f L = 2\pi \times 50 \times 95.5 \times 10^{-3}$$
$$= 30\,\Omega$$

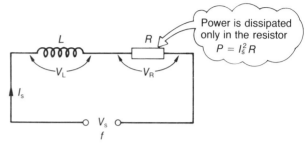

I_s and V_s are r.m.s. values.

(a) The circuit

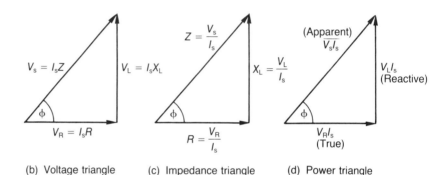

(b) Voltage triangle   (c) Impedance triangle   (d) Power triangle

Figure 4.14 *The voltage, impedance and power triangles*

ϕ is the circuit phase angle; it is the angle by which the supply current, I_s, lags the supply voltage, V_s.

The impedance of the circuit is given by Equation 4.13:

$$Z = \sqrt{(R^2 + X_L^2)} = \sqrt{(40^2 + 30^2)} = 50\,\Omega$$

The current drawn from the supply is:

$$I_S = \frac{V_S}{Z}$$
$$= \frac{250\,V}{50\,\Omega}$$
$$= 5\,A$$

Example 4.6 *For the circuit shown in Figure 4.15, calculate the voltage across the inductance and that across the resistance. Also, estimate the power dissipated by the circuit.*

From the previous example we know that $I_S = 5\,A$ and that $X_L = 30\,\Omega$.
The voltage across the inductance is:

$$V_L = X_L I_S = 30 \times 5$$
$$= 150\,V$$

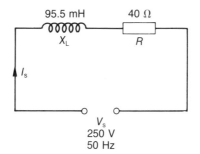

Figure 4.15 *Circuit for Examples 4.5, 4.6 and 4.7*

The voltage across the resistance is:

$$V_R = RI_s = 40 \times 5$$
$$= 200\,\text{V}$$

All of the power dissipated by the circuit is in the resistor alone – an inductance does not dissipate power. Power in the resistor is:

$$I_S^2 R = 5^2 \times 40 = 1000\,\text{W}$$

Alternatively, this could have been calculated from:

$$\frac{V_R^2}{R} = \frac{200^2}{40} = 1000\,\text{W}$$

Example 4.7   *For the circuit in Figure 4.15, calculate:*

(a)   the circuit phase angle
(b)   the apparent power drawn from the supply
(c)   the true power dissipated.

(a)   From the voltage triangle in Figure 4.16, we can calculate the phase angle ϕ from $\tan\phi = V_L/V_R$ or $\sin\phi = V_L/V_S$ or $\cos\phi = V_R/V_S$. The

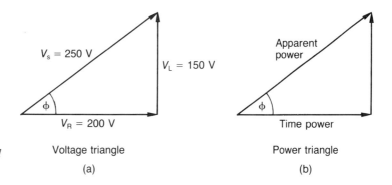

Figure 4.16 *Voltage triangle and power triangle for Example 4.7*

value of $V_S = 250$ V is given on the circuit and in Example 4.6 above; the values of $V_R = 200$ V and $V_L = 150$ V were calculated. From any one of these trigonometrical ratios we can calculate ϕ as 36.86°.

(b)   Apparent power drawn = supply volts × supply current.

The supply volts are given as 250 V and we calculated I_S as 5 A in Example 4.5.

Therefore, the apparent power is:

$$V_S \times I_S = 250 \times 5 = 1250 \text{ W}$$

(c)   The true power dissipated was calculated in Example 4.6 as 1000 W. However, had we not known this, we could have used Figure 4.16(a) to determine $\phi = 36.86°$ and then, putting this value of ϕ into Figure 4.16(b), we could now calculate the true power from:

$$\text{True power} = \text{apparent power} \times \cos \phi$$
$$= 1250 \times \cos 36.86° = 1000 \text{ W}$$

Cos ϕ is called the circuit *power factor*.

4.5.2 Series capacitance and resistance

The circuit for this combination is shown in Figure 4.17. Since the voltage across a capacitor cannot be changed immediately, it must lag any changes made to the current flowing into it. Thus, with an assumed anticlockwise rotation of the phasors, the common current I_S is drawn horizontally, left to right, and the lagging capacitor voltage, V_C, is drawn vertically 'down'. Once again, the supply voltage, V_S, is the vector sum of V_R and V_C, but this time the supply current, I_S, is leading by an angle ϕ. The formula for calculating the circuit impedance is similar to that for the series inductance resistance.

Example 4.8   *Calculate the phase angle and current in a circuit consisting of a 12 Ω resistor in series with a 31.8 μF all connected across a 10V, 1000 Hz supply. What is the value of the power factor?*

The phase angle, ϕ, can be obtained from any one of the voltage, impedance or power triangles. The easiest one to use in this case is the impedance triangle. The question gives the value of R and the means for calculating the value of the reactance X_L. Figure 4.17(c) shows the impedance triangle for the capacitive series circuit.

$$X_C = \frac{1}{2\pi f C} = \frac{1}{2\pi \times 1000 \times 31.8 \times 10^{-6}} = 5\,\Omega$$

From the impedance triangle, $\phi = \tan^{-1} X_C / R = \tan^{-1} 5/12$:

$$\phi = 22.62°$$

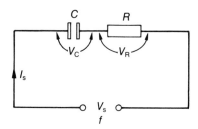

(a) The circuit

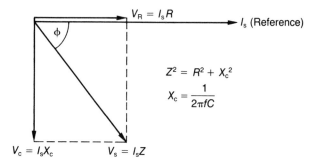

(b) The voltage phasor diagram

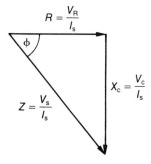

Figure 4.17 *Series capacitance and resistance*

(c) The impedance triangle.

In order to determine the current in the circuit, we need to know its impedance, Z:

$$Z = \sqrt{(R^2 + X_C^2)} = \sqrt{(12^2 + 5^2)} = 13\,\Omega$$

The circuit current, I_S, is given by:

$$I_S = \frac{V_S}{Z} = \frac{10}{13} = 0.77\,A$$

Because the power triangle is mathematically similar to the impedance triangle we can deduce that:

$$\text{Power factor} = \frac{\text{true power}}{\text{apparent power}}$$

$$= \text{same ratio as } \frac{R}{Z}.$$

Therefore, power factor $= 12/13 = 0.92$.

4.5.3 Inductance, capacitance and resistance in series

Figure 4.18(a) shows this arrangement. The various equations and phasor diagrams must now take into account that there are two opposing types of reactance involved and that it is often their difference or net reactance which has to be used. This aspect is illustrated by phasor diagrams in Figures 4.18(b) and (c). Whether the circuit phase angle, ϕ, is lagging or leading depends upon the relative proportions of X_L and X_C.

Example 4.9   *The circuit shown in Figure 4.18(a) has a resistor of 10 Ω, an inductor of 0.1 H, a 50 μF capacitor and a 100 V, 50 Hz supply. Calculate the circuit current, the phase angle and the power factor.*

The first step is to ascertain the two reactances, X_L and X_C, and then the total impedance, Z, of the circuit:

$$X_L = 2\pi fL = 2\pi \times 50 \times 0.1 = 31.42 \, \Omega$$

$$X_C = \frac{1}{2\pi fC} = \frac{10^6}{2\pi \times 50 \times 50} = 63.7 \, \Omega$$

The net reactance, X_{NET}, is given by:

$$X_{NET} = X_C - X_L = 63.7 - 31.42 = 32.28 \, \Omega$$

$$Z = \sqrt{(R^2 + X_{NET}^2)} = \sqrt{(10^2 + 32.28^2)} = 33.8 \, \Omega$$

The circuit current $= V_S/Z = 100/33.8 = 2.96 \, \text{A}$.
The circuit phase angle, ϕ, is given by:

$$\phi = \tan^{-1}\frac{X_{NET}}{R} = \tan^{-1}\frac{32.28}{10} = \tan^{-1}3.228 = 72.8°$$

The power factor can be obtained directly from the cosine of the phase angle:

Power factor $= \cos\phi = \cos 72.8° = 0.3 \, \text{A}$

4.5.4 Series resonance

The phasor diagrams of Figure 4.18 become much simpler if the inductive and capacitive reactances are equal in value. If this is the case, the common series circuit current flowing through them causes their equal but opposite voltage drops to cancel each other. As far as the supply

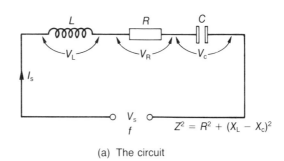

(a) The circuit

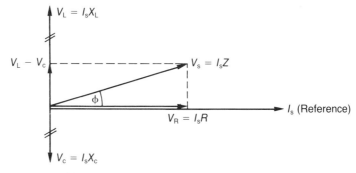

(b) The phasor diagram for the case where $X_L > X_c$

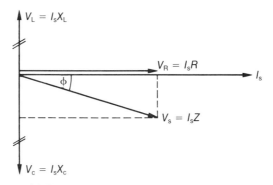

Figure 4.18 *Series inductance, capacitance and resistance*

(c) Phasor diagram for the case where $X_c > X_L$

voltage generator is concerned, the inductance and the capacitance have been effectively shorted out, leaving only the resistance, R, to impede the flow of current:

$Z = R$ at resonance

The phasor diagram for the resonant series circuit is given in Figure 4.19. The resonant condition can be arranged either by selecting a

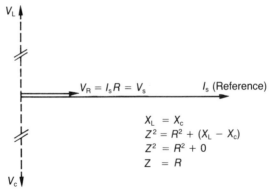

At resonance, $V_L = V_c$ and cancel each other
leaving only $V_R = V_s$

Series resonance occurs at one particular
frequency, f_o, which makes $X_L = X_c$

Therefore $2\pi f_o L = \dfrac{1}{2\pi f_o C}$

$$f_o = \frac{1}{2\pi\sqrt{LC}}\ \text{Hz}$$

Figure 4.19 *Series resonance*

matching inductance and capacitance or by adjusting the supply voltage
frequency to make $X_L = X_C$.

The frequency which will cause the series *LCR* (inductance, capacitance and resistance) to resonate is obtained by transposing the equation:

$$X_L = X_C$$

Therefore, $2\pi f L = 1/(2\pi f C)$ and making f the subject of the equation
we have:

$$f = \frac{1}{2\pi\sqrt{(LC)}}\,\text{Hz} \qquad\qquad [4.15]$$

At resonance, the series circuit assumes several important conditions.
These are apparent from a careful look at and understanding of the
phasor diagrams in Figures 4.18 and 4.19:

- $Z = R$
- The circuit is purely resistive
- $V_R = V_S$
- Z is at its minimum value
- I_S is at its maximum value
- The phase angle, ϕ, between I_S and V_S is zero
- The power factor is unity
- $V_L = V_C$
- $X_L = X_C$

Examination questions on series LCR circuits very often indicate that the circuit is in resonance only by saying that one or other of the above conditions applies.

Example 4.10 *A series LCR circuit, as shown by Figure 4.18(a), has an inductance of 50 mH, a resistance of 10 Ω and a capacitance of 20μF. The frequency of the 20 V supply is adjusted until the circuit current is at its maximum. Calculate:*

(a)   the frequency of the supply
(b)   the impedance of the circuit
(c)   the current flowing
(d)   the volt drop across the inductance
(e)   the voltage across the capacitance
(f)   the voltage across the resistance.

The fact that the circuit is passing its maximum current indicates that it is in resonance.

(a)   The frequency of resonance can be calculated using Equation 4.15:

$$\sqrt{(LC)} = \sqrt{(50 \times 10^{-3} \times 20 \times 10^{-6})} = 10^{-3}$$

$$f = 1/(2\pi \times \sqrt{(LC)}) = 1/(2\pi \times 10^{-3}) = 159\,\text{Hz}$$

(b)   The impedance of the circuit at resonance is simply R, because the net reactance is zero.
    Therefore we can say, $Z = R = 10\,\Omega$.
(c)   The circuit current, $I_S = V_S/Z = 20/10 = 2\,\text{A}$
(d) $X_L = 2\pi fL = 2\pi \times 159 \times 50 \times 10^{-3} = 50\,\Omega$
    $V_L = I_S \times X_L = 2 \times 50 = 100\,\text{V}$
(e)   At resonance, $V_C = V_L = 100\,\text{V}$.
(f) $V_R = I_S \times R = 2 \times 10 = 20\,\text{V}$.

It is interesting to note how this circuit can act as an amplifier. The voltage across the inductance is 100 V, five times that of the 20 V supply voltage.

4.5.5 Power factor correction

In Section 4.5.2, we saw how the phase angle between the current (I_S) flowing through and the voltage (V_S) across a component affects the power developed by that component. The apparent power supplied to the component is $(V_S \times I_S)$ volt-amps, but the true power developed is the voltage multiplied by only the component of the current that is in phase with it, that is $(V_S \times I_S \cos\phi)$ watts. This means that an inductive load must be supplied with current through cables of sufficient size to carry I_S yet only $I_S \cos\phi$ is being converted into useful power.

If the load is sufficiently large, perhaps a factory with many induction motors driving machine tools, the current lag caused may be so large that the Electricity Board is obliged to increase their charges. Otherwise, the Board is paying to produce $I_S \times V_S$ electrical power but charging the factory for only $V_S \times I_S \cos \phi$.

The additional charges can be avoided if the factory manager invests in equipment to correct the factory power factor. If the power factor, $\cos \phi$, is less than a certain value, say 0.8, it may become an economical proposition to take steps to reduce ϕ to zero. This is done by connecting the motor which is taking a lagging current in parallel with a capacitor which takes a leading current. Consider the following example.

Example 4.10   *A 240 V single phase motor takes a current of 50 A at a 0.8 lagging power factor. Calculate the value of the capacitor which must be connected in parallel with the motor terminals in order to increase its power factor to unity.*

The circuit diagram and the voltage phasor diagram for the basic motor are shown in Figures 4.20(a) and (b). Cos ϕ is the power factor, given as 0.8, so the angle ϕ is 36.89°. It is this angle which must be reduced to zero if the power factor is to be increased to unity. With $\cos \phi = 1$, the current drawn from the supply will be in phase with the supply voltage and the circuit as a whole will appear to the supply as a pure resistor.

Figure 4.20(c) illustrates the problem of reducing the angle ϕ to 0°. The voltage phasor, V_S, and the current, I, in Figure 4.20(b) have been extracted and redrawn in Figure 4.20(c) but with V_S providing the reference direction. Also, the supply current, I, has been resolved into its horizontal and vertical components, $I \cos \phi$ and $I \sin \phi$ respectively. We must be quite clear as to what these current phasors represent.

- I is both the supply current and the current that flows through the motor to produce the mechanical output power.
- $I \cos \phi$ is the proportion of I that is converted into mechanical power, assuming 100% conversion efficiency. $I \cos \phi$ is the current component that is in phase with the voltage, V_S, across the motor.
- $I \sin \phi$ is the quadrature current component that produces no power and is responsible for swinging I out of phase with V_S by the lagging angle ϕ. If the angle ϕ is to be reduced to zero, it is this current phasor that must be removed.

In our particular example, I is given as 50 A. With the power factor of 0.8, and V_S equal to 240 V, the power developed by the motor is:

$$V_S \times I \cos \phi = 240 \times 50 \times 0.8 = 9600 \, \text{W}$$

From this we can see that the in-phase current component is 40 A and this must be maintained if the motor is to produce its required mechanical output. For this to be so, we must continue to feed 50 A to the motor but if we connect a capacitor in parallel with the motor we can use it

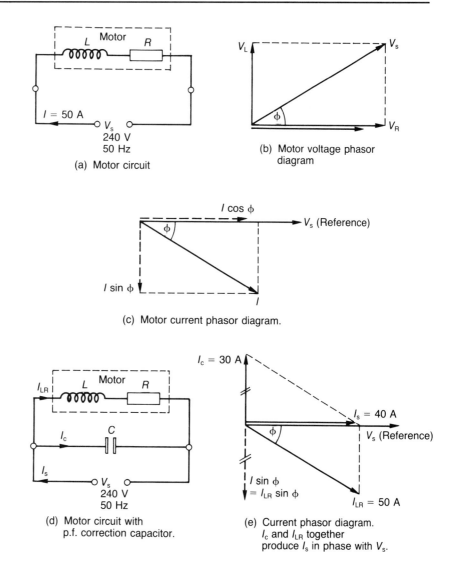

Figure 4.20 *Power factor correction*

(a) Motor circuit

(b) Motor voltage phasor diagram

(c) Motor current phasor diagram.

(d) Motor circuit with p.f. correction capacitor.

(e) Current phasor diagram. I_c and I_{LR} together produce I_s in phase with V_s.

effectively to supply part of that 50 A. This is possible because a capacitor takes a leading current as opposed to the lagging current drawn by the motor. If we choose the correct value of capacitor, the leading and lagging currents cancel. This situation is shown by Figures 4.20(d) and (e). Note how the capacitor current, I_C, is arranged to be equal but opposite in direction to the original $I \sin \phi$. With the capacitor fitted, the original supply current, I, now becomes I_S, the vector sum of I_{LR} to the motor and I_C to the capacitor:

$$I_C = 50 \times \sin 36.89° = 30 \, \text{A}$$

We can now calculate the value of capacitor required to pass this current.

The capacitive reactance, $X_C = V_S/I_C = 240/30 = 8\,\Omega$. Therefore:

$1/(2\pi fC) = 8\,\Omega$

$C = 1/(2\pi \times 50\,\text{Hz} \times 8\,\Omega) = 397.9\,\mu\text{F}$

4.5.6 Parallel resonance

The principle of parallel resonance has been covered in the previous section. The correction of the lagging power factor of the inductive motor entailed bringing the supply current and the supply voltage into phase, reducing the phase angle to zero degrees. It will have been noticed how the original supply current of 50 A was reduced to one of 40 A when I_S and V_S became in-phase. This, in fact, is a case of parallel resonance where there is *current amplification*. The current in the motor-capacitor loop is greater than that supplied.

This compares with the *voltage amplification* of the series resonant circuit where the voltages across the inductance and capacitance were greater than the supply voltage.

However, unlike the series resonant case, parallel resonance does not occur when $X_L = X_C$. The criterion for parallel resonance is definitely where $\phi = 0°$. If the coil in the circuit has X_L at least ten times larger than R, then the same formula (Equation 4.15) for calculating the resonant frequency applies. Again, unlike the series resonant circuit, it has a maximum impedance at the resonant frequency rather than a minimum.

The parallel resonant circuit is much used in electronic oscillator, radio and filter circuits but these are outside the scope of this book.

Exercises

4.1 A 10 V, 1 kHz supply is connected to the following inductances. In each case calculate the current drawn from the supply:
(a) 2 mH
(b) 100 mH
(c) 1 H.

4.2 A 10 V, 1 kHz supply is connected to the following capacitances. In each case calculate the current flowing:
(a) 1 mF
(b) 2 μF
(c) 5 nF.

4.3 A pure inductance of 1.25 mH is connected in series with a pure resistance of 27 Ω. If the frequency of the 10 V, sinusoidal supply is 5 kHz, calculate:
(a) the impedance of the circuit
(b) the current flowing
(c) the power dissipated.

4.4 A coil of inductance 160 mH and resistance 22 Ω is connected in series with a 50 Ω resistor to a 230 V, 50 Hz supply. Calculate:
   (a) the impedance of the circuit
   (b) the current in the circuit
   (c) the circuit phase angle
   (d) the volt drop across the 50 Ω resistor
   (e) the volt drop across the coil.

4.5 A capacitor of value 65 μF is connected in series with a 12 Ω resistor across a 50 V, 100 Hz supply. Calculate:
   (a) the circuit impedance
   (b) the current flowing
   (c) the circuit phase angle
   (d) the power factor.

4.6 An inductance of 150 mH, a resistance of 68 Ω and a 10 μF capacitor are all connected in series across a 450 V, 200 Hz supply. Calculate:
   (a) the impedance of the circuit
   (b) the current flowing
   (c) the apparent power supplied
   (d) the true power dissipated.

4.7 A coil having a resistance of 22 Ω and an inductance of 75 mH is connected in series with a 50 μF capacitance to a 25 V supply. If the frequency of the supply is adjusted until maximum current is flowing in the circuit, calculate:
   (a) the frequency
   (b) the circuit impedance
   (c) the current flowing.

4.8 The current in a series resonant circuit is 180 μA. If the applied voltage is 6 mV at a frequency of 100 kHz and the circuit inductance is 45 mH, calculate:
   (a) the circuit resistance
   (b) the circuit capacitance.

4.9 A coil of resistance 40 Ω, 160 mH inductance is connected in parallel with a 25 μF capacitor across a 250 V, 60 Hz supply. Calculate:
   (a) the current in each branch
   (b) the supply current
   (c) the circuit phase angle
   (d) the circuit impedance
   (e) the power factor.

4.10 An inductive load comprising a 0.2 H inductance in series with a 40 Ω resistance is connected across a 230 V, 50 Hz mains supply.
   (a) Calculate the current drawn by the load and its power factor.
   (b) If the power factor is increased to unity, calculate the new current drawn from the mains supply and the value of the capacitance which must be connected in parallel across the load terminals to achieve this new power factor.

5 Three-phase currents

5.1 Electrical power distribution

In Chapter 2 we saw how electrical power can be obtained from a number of different sources. Batteries tend to be used for the storage of electricity and its supply at low voltage (less than 100 V) d.c. Mains operated power supply units (PSUs), in general, provide a low voltage d.c. output. Rotating machines in the form of mechanically driven generators can produce direct current at pressures as high as kV if required. Driven rotating machines in the form of single-phase or three-phase alternators are the main courses of a.c. power. The latter machine is the basic source of electrical energy distributed through the *National Grid System*.

5.1.1 The need for the National Grid System

Until the early 1930s the supply of electrical power throughout the country was in the hands of local electrical generating and distribution companies. There was no national standardization of voltage and both d.c. and a.c. systems co-existed. Clearly this was a very restricted and inflexible arrangement and the National Grid System was established to provide a better common system for all consumers. Economics demanded that the power stations which supplied the grid distribution network should be built either near to a convenient fuel supply or near to a major user of electricity. The electrical power had to be carried over large distances and the cheapest system was, and still is, an overhead power line suspended from pylons.

Some of the immediate advantages of having a common national electrical system are:

- All electrical appliances can be standardized to work from same national supply.
- Power stations can be switched in or out quite quickly to match generating capacity and consumer demand.
- Overloaded power stations can be assisted by those under a light load.
- A multi-path grid distribution system will help to ensure continuity of supply in the event of a single line failure.

5.1.2 Factors affecting the specification of the electrical supply

High voltage

Even though overhead cables may be the cheapest method of distributing electrical power it is still very expensive. The steel cored (for high tensile strength), copper or aluminium (for low electrical resistivity) cables need to be as narrow gauge as possible in order to reduce their cost (bought by the tonne), to make their handling during erection easier (cheaper) and to lessen the required strength of the supporting pylons (cheaper). It is the current handling requirement of the cable which determines its size and cost. For the transmission of a given amount of electrical power the way to restrict the size of cable is to use a high voltage. For example, in order to transmit 10 kW to a distant point, even assuming no resistive losses, a 100 V system would need a cable current-carrying capacity of 10 000 W/100 V = 100 A. If the system voltage were increased to 20 000 V then the current need be only 10 000 W/20 000 V = 0.5 A and the cable diameter can be much reduced. This then is a very good reason for choosing a very high system voltage.

Alternating current

Having decided that the system voltage should be very high for power transmission purposes, we then have to consider the problems associated with generating and using these very high voltages, which can be as high as 400 kV. The ideal situation is to generate the electrical power at a low voltage and then increase or *step-up* the voltage for distribution over long cables. Since we would not be happy with the safety aspect of connecting 400 kV to our domestic appliances, we must also have a method of reducing or *stepping-down* the voltage for the consumer to use. At the present time the consumer receives a domestic supply at 240 V and light industry at 415 V.

The ideal device for changing voltages is the *transformer*. These are a.c. devices and were described in Chapter 1. Therefore, in order to use transformers in the system we must use alternating currents (a.c.) rather than direct currents (d.c.). The frequency of the a.c. is 50 Hz and it is customary to describe the single-phase and three-phase voltages respectively as follows:

240 V/1-ph/50 Hz or simply 240/1 ϕ/50
415 V/3-ph/50 Hz or simply 415/3 ϕ/50

Three-phase supply

With an a.c. system there is the choice of using single-phase or three-phase currents. The latter was selected as the most suitable for the following reasons:

- For given electric motor dimensions, a three-phase machine will self-start when power is applied, it will run with less vibration (compare a six-cylinder car engine to that of a single-cylinder motor cycle engine), and it develops more mechanical output power. Further, three-phase induction motors have only one moving part (the rotor) with no brush gear connections. They are therefore cheap, reliable and easy to maintain.

- As we now know, the instantaneous pattern of currents in the star-connected phase windings of the alternator is produced by the rotating magnetic field within the alternator. This same changing pattern of alternator phase currents is passed to any motor connected to it. A three-phase induction motor also has star-connected windings which therefore set up a rotating magnetic field within the motor around the rotor which is pulled around with it. Hence the self-starting effect of the three-phase motor.

- Two voltages are readily available: $415/3\phi/50$ or $240/1\phi/50$ (see Figure 5.1).

5.1.3 The national distribution system

Figure 5.1 shows the general idea of the National Grid System. Typical power station alternator voltages are in the order of 20 kV and these are stepped-up through large three-phase transformers to feed sections of transmission lines which may be working at 132 kV, 275 kV or even 400 kV. Heavy, medium and light industrial consumers may tap-off their power requirements at 33 kV or 11 kV; if they do, they obtain a cheaper tariff. Very small industrial concerns take power at 415 V/3-ph/50 Hz using a four-wire cable. This is ideal for driving the motors fitted to lathes, grinders, milling machines and the like and can easily supply the 240 V/1-ph/50 Hz required for office use. Similarly, the local substations supplying domestic users take a four-wire 415 V, three-phase supply from which 240 V, single-phase domestic services are tapped-off. This latter is achieved by tapping across any one of the red, yellow or blue phase lines and the neutral line. Local electrical distribution engineers need to keep balanced the current drawn from each of the three phases of the local substation.

5.2 The relationship between line and phase voltages

Figure 2.12, in Chapter 2, shows the general $120°$ layout of the three-phase alternator windings and if we draw the three phasors to represent the three r.m.s. phase voltages we see the result shown in Figure 5.2(a). For example, V_{NR} is the red phase voltage from N to R; similarly, V_{NY} and V_{NB} are the yellow and blue phase voltages respectively. The line voltage between the red and yellow phases is seen to be made up from the two individual phase voltages concerned. If, for example, we were to connect a resistor between points R and Y, the current which would

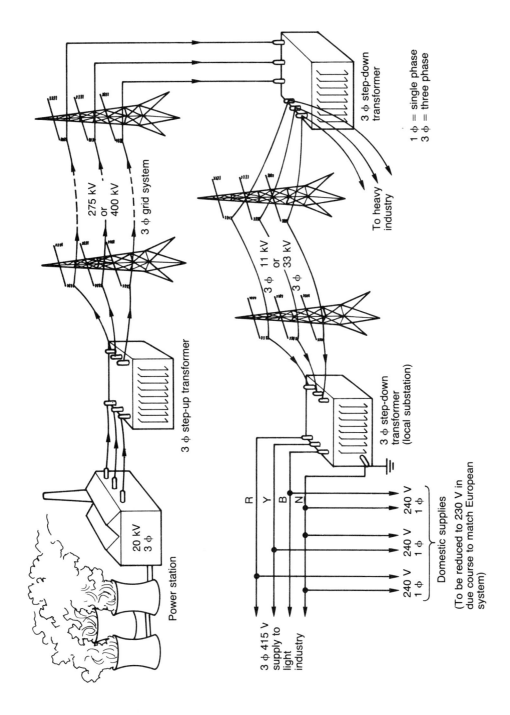

Power station

20 kV
3 φ

3 φ step-up transformer

275 kV
or
400 kV

3 φ grid system

3 φ step-down transformer

1 φ = single phase
3 φ = three phase

To heavy industry

3 φ 11 kV
or
33 kV
3 φ

3 φ step-down transformer (local substation)

R
Y
B
N

3 φ 415 V
supply to
light
industry

240 V
1 φ
240 V
1 φ
240 V
1 φ
240 V
1 φ

Domestic supplies

(To be reduced to 230 V in
due course to match European
system)

Fire 5.1 *Electrical power distribution system*

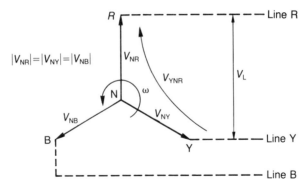

(a) Phasor diagram showing three phase voltages
each separated by 120°

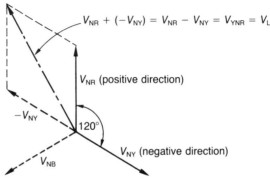

$$V_{NR} + (-V_{NY}) = V_{NR} - V_{NY} = V_{YNR} = V_L$$

(b) Phasor diagram to show how V_L is the
difference voltage between two phases.

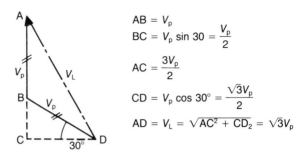

$AB = V_p$

$BC = V_p \sin 30 = \dfrac{V_p}{2}$

$AC = \dfrac{3V_p}{2}$

$CD = V_p \cos 30° = \dfrac{\sqrt{3}V_p}{2}$

$AD = V_L = \sqrt{AC^2 + CD_2} = \sqrt{3}V_p$

*Figure 5.2 Relationship between
line and phase voltages*

(c) Line voltage is $\sqrt{3}$ times phase voltage.

flow through the resistor would be that driven by the voltage difference
between the ends of the resistor, namely, the voltage *phasor difference*
between the points R and Y. Thus, V_{YNR}, being the line voltage V_L, is the
phasor difference between V_{NR} and V_{NY}. In fact we add the phasor

negative of V_{NY} to V_{NR} to obtain V_{YNR}. Figure 5.2(b) shows the graphics of the phasor differencing operation and Figure 5.2 (c) shows how the magnitude of the r.m.s. line voltage is deduced as being $\sqrt{3}$ times that of the r.m.s. phase voltage.

5.3 Star- and delta-connected loads

5.3.1 Star loads

Figure 5.3 shows our now familiar three-phase alternator (or local transformer output winding) with the three phase-windings connected centrally at the star point. This is called a *star connection* and in the previous section we deduced the $\sqrt{3}$ relationship between its phase and line voltages. If we now connect a similar star-connected resistive load to the alternator or transformer we shall have a flow of phase and line currents. If the load has the same value of resistor in each phase, the three r.m.s. phase currents drawn will be identical and we have what is known as a *balanced load*.

Note that while the phase voltage across each of the phase resistors is less than the line voltage (because the line voltage itself is expended across two phase resistors), the line currents and the phase currents are one and the same. (They must be the same because where else can the line current flow other than through the phase load to which it is connected?) Also

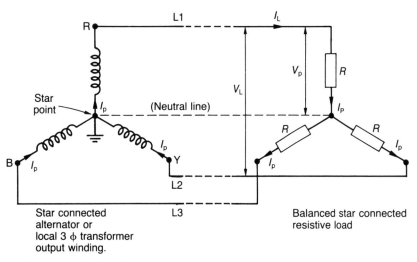

Star connected
alternator or
local 3 φ transformer
output winding.

Balanced star connected
resistive load

V_p = r.m.s. phase voltage
V_L = r.m.s. line voltage between any pair of lines
I_p = r.m.s. phase current
I_L = r.m.s. line current

For a balanced star-connected load, the r.m.s. phase voltages and currents are of the same magnitude as are the three r.m.s. line voltages and currents.

Also, $\sqrt{3}\, V_p = V_L$ and $I_p = I_L$

Figure 5.3 *Star-connected balanced resistive load*

remember that we are dealing with r.m.s. values of voltage and current and while these are conveniently constant, the instantaneous phase and line voltages and currents are ever-changing sinusoidal quantities.

For the balanced star-connected load we can say:

$$\sqrt{3}V_P = V_L \tag{5.1}$$

and

$$I_P = I_L \tag{5.2}$$

5.3.2 Delta (or mesh) loads

Figure 5.4 shows the electrical diagram for a three-phase source driving a balanced three-phase load with its three equal value resistors connected in a delta configuration. Analysis of the circuit follows a similar process to that in the previous section for the star load. The results are much the same in that the relationship between the phase and line voltages and currents is the same factor of $\sqrt{3}$ but with a slightly different twist. This time, as can be seen directly from the arrows marking the current flow, the line current divides itself between two phases and therefore the phase currents must be smaller than the line current from which they are formed. The individual phase currents are not half of the line current but $1/\sqrt{3}$ times the line current. But the phase voltage and the line voltage are one and the same: this is apparent just from visual inspection of the circuit.

For the balanced delta-connected load we can say:

$$V_P = V_L \tag{5.3}$$

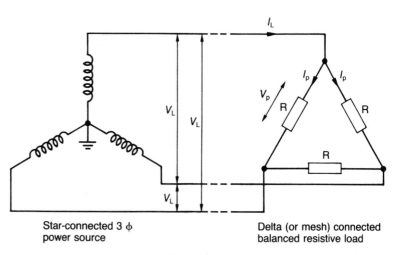

Star-connected 3 φ
power source

Delta (or mesh) connected
balanced resistive load

$$V_L = V_p \qquad \sqrt{3}\,I_p = I_L$$

Figure 5.4 *Delta (or mesh) connected balanced resistive load*

and

$$\sqrt{3}I_P = I_L \qquad\qquad\qquad [5.4]$$

5.4 Three-phase power

5.4.1 Calculation of three-phase load power

One of the major differences between the star and delta connections is in the power they dissipate. The power for both connections when the loads are balanced is determined simply by finding the power in one load element and multiplying it by three for the full load. The power in any one of the phase resistors, R, is given by:

Phase power $= (I_P)^2 R$ watts $\qquad\qquad\qquad [5.5]$

For the three equal resistor elements, the power dissipated must be three times this; just as the heat emitted by three heating elements would be three times that emitted by one element. Therefore we can say:

Total load power $= 3(I_P)^2 R$ watts $\qquad\qquad\qquad [5.6]$

Since we know the relationship between the phase and line voltages and currents for both the star and delta connections, it can be shown that for both cases the total power dissipated in terms of the line, rather than the phase, quantities is:

$$\sqrt{3}V_L I_L \cos\phi \text{ watts} \qquad\qquad\qquad [5.7]$$

where ϕ is the phase angle between I_P and V_P, and $\cos\phi$ is the power factor.

Remember that this applies whether the total load is star-connected or delta-connected.

5.4.2 Measurement of three-phase load power

It can be seen from Equation 5.7 that the power dissipated in a three-phase system can be determined from knowing the value of the line voltage and currents and the power factor, $\cos\phi$. A wattmeter is a cleverly designed power meter which takes all these factors into account when producing a power reading. Very briefly, the wattmeter has a fixed coil through which the line *current* is made to flow. Another coil, this one pivoted to allow it to swing, carries a very small current which is proportional to the line *voltage*. Both the fixed and moving coil currents produce magnetic fields which interact to turn the moving coil against the resistive force of a spring. Fixed to the moving coil is a pointer which moves over a scale calibrated in watts and this indicates the measured power.

If the load is balanced, only one wattmer need be used to measure the power in a single phase; the wattmeter reading is then multiplied by three to obtain the total load power dissipation. Figure 5.5 shows the connections required for power measurement in balanced star- and delta-connected loads. Should there be a possibility of the three load elements being unbalanced then two wattmeters will be required – Figure 5.6 shows how they are connected. While Figure 5.6 shows a star-connected load, it could equally well have been delta-connected.

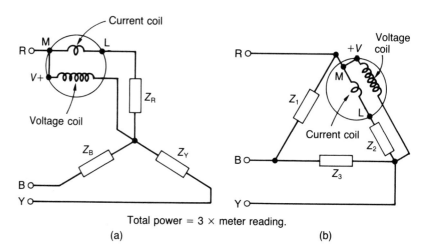

Figure 5.5 *Single wattmeter connection for power measurement for balanced (a) star-connected and (b) delta-connected loads*

Total power = 3 × meter reading.

(a)   (b)

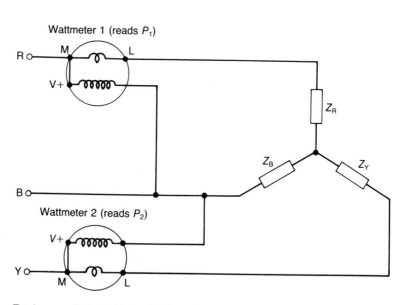

Figure 5.6 *Two wattmeters must be used if there is a possibility of load unbalance*

Total power dissipated in load is $P_1 + P_2$

5.5 Worked three-phase problems

Example 5.1   *A balanced star-connected resistive three-phase load comprises 25 Ω resistors and is connected to a 398 V three-phase supply. Calculate:*
(a) the voltage across each of the load resistors
(b) the current flowing through each load resistor
(c) the current in each supply line.

The circuit is shown by Figure 5.3 and we are required to determine (a) V_P, (b) I_P and (c) I_L. Since the load resistors are all of the same value, the load is balanced and we know therefore that the same r.m.s. voltages and currents must exist in each limb of the three-phase load. Thus, we need to concern ourselves with calculations for only one phase; the other two will have the same results.

Now a 398 V three-phase supply means that V_L has an r.m.s. value of 398 V and so 398 V exists between any pair of the three supply lines, each of which is carrying an r.m.s. current of I_L. We further know that for a star-connected load, since V_L is spread across *two* phase loads, V_P must be *less* than V_L. The conversion factor between V_P and V_L is $\sqrt{3}$ so to make V_P less than V_L we must divide V_L by $\sqrt{3}$.

(a) $V_P = V_L/\sqrt{3} = 398\sqrt{3} = 229.79$ V.
(b)   Now, I_P is the phase current driven through each of the 25 Ω phase load resistors by the phase voltage, V_P. Therefore we can use Ohm's law and write $V_P = V_P/R_P$ where R_P is the phase resistance.
   Therefore, $I_P = 229.79/25 = 9.19$ A.
(c)   For a star-connected load we know that $I_P = I_L$. Therefore we can say that $I_L = 9.19$ A.

Example 5.2   *Suppose that the situation is the same as for Example 5.1 except that the balanced three-phase resistive load is connected in a delta configuration.*

The circuit is shown in Figure 5.4. We know that for a delta load, the line voltages and the phase voltages are the same: $V_L = V_P$. Clearly the line current, I_L, is shared between two load phases and therefore I_P is *less* than I_L. The conversion factor, as ever, with these three-phase problems is $\sqrt{3}$ and so we must multiply I_P by $\sqrt{3}$ to obtain the value of I_L.

(a) $V_P = V_L = 398$ V.
(b) $I_P = V_P/R_P = 398/25 = 15.92$ A.
(c) $I_L = \sqrt{3}V_P = \sqrt{3} \times 15.92 = 27.57$ A.

Example 5.3   *Three 30 Ω load resistors are connected (a) in star and (b) in delta to a 415 V three-phase supply. For each connection calculate the total load power dissipation and state which is the larger.*

(a)   *Star-connected load*

$V_P = V_L/\sqrt{3} = 415/\sqrt{3} = 239.6\,V$

$I_P = V_P/R_P = 239.6/30 = 7.99\,A$

The power dissipated in a single phase is:

$V_P = I_P = 239.6 \times 7.99 = 1914.4\,W$

The power dissipated in all three phases is three times that dissipated in one, that is:

$3 \times 1914.4 = 5743.2\,W$.

(b)   *Delta-connected load*

$V_P = V_L = 415\,V$

$I_P = V_P = 415/30 = 13.83\,A$

The power dissipated in a single phase is:

$V_P = I_P = 415 \times 13.83 = 5739.45\,W$

The power dissipated in all three phases is three times that dissipated in one, that is:

$3 \times 5739.45 = 17\,218.35\,W$

Notice how *the power dissipated in the delta connection is three times that of the star connection.* Delta connections are the more powerful but draw more current from the mains supply. Now, if an electric motor is started from rest, the starting current it draws is in the order of six times its normal full-load running current. If the motor is sufficiently large, this heavy initial current drain can cause the local supply voltage to momentarily fall. This voltage dip can have undesirable effects on other local mains-driven equipment, especially equipment which is particularly voltage sensitive. For this reason, large a.c. induction motors have their stator windings switchable between the star and delta modes. The changeover action can be achieved automatically, using a centrifugal switch, or manually by means of a *star–delta starter.* The motors are *star*ted in the *star* connection but once run up to full speed they are switched to delta. The star connection draws minimum current while starting and the delta connection allows the motor to develop its maximum output power under load.

Example 5.4   *A star-connected load consists of three identical coils each of resistance 25 Ω and inductance 120 mH. If the line current is 5.5 A and the supply frequency 50 Hz, calculate:*
(a) the line voltage
(b) the power dissipated by the load.

It always helps to first sketch the circuit and mark on it the known quantities – see Figure 5.7. The method of tackling this problem is first

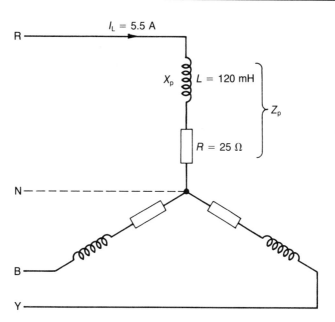

Figure 5.7 *Three-phase balanced star-connected load for Example 5.4*

to calculate the phase impedance (since the load is balanced, all phases will have the same impedance), and knowing the phase current (it is the same as the given line current for this star-connection), we can use Ohm's law to determine the phase voltage. The line voltage is $\sqrt{3}$ times the phase voltage and so we have the answer to (a).

For (b), the power in one phase is the (phase current)^2 × the phase resistance. *Note that the phase inductance, L, does not dissipate power; only the resistive element, R, dissipates power as heat.*

Now let us try the calculations:

(a) $X_P = 2\pi f L = 2 \times \pi \times 50 \times 120 \times 10^{-3} = 37.7\,\Omega$
$R = 25\,\Omega$
$Z_P = \sqrt{(R^2 + X_P^2)} = \sqrt{(25^2 + 37.7^2)} = 45.24\,\Omega$
$\therefore V_P = I_P Z_P = 5.5 \times 45.24 = 248.8\,\text{V}$
$\therefore V_L = \sqrt{3} \times 248.8 = 431\,\text{V}$

(b)   Power dissipated per phase is $I_P^2 R = 5.5^2 \times 25$
$= 756.25\,\text{W}$
$\therefore$ Total load power dissipated $= 3 \times 756.25$
$= 2269\,\text{W}$

Example 5.5   *Three 50 μF capacitors are connected (a) in star and (b) in delta to a 390 V, three-phase, 50 Hz supply. Calculate the line current drawn for each connection.*

The first step is to sketch the relevant circuit diagram and mark on it the known facts – see Figure 5.8. Now we carry out the calculations as follows:

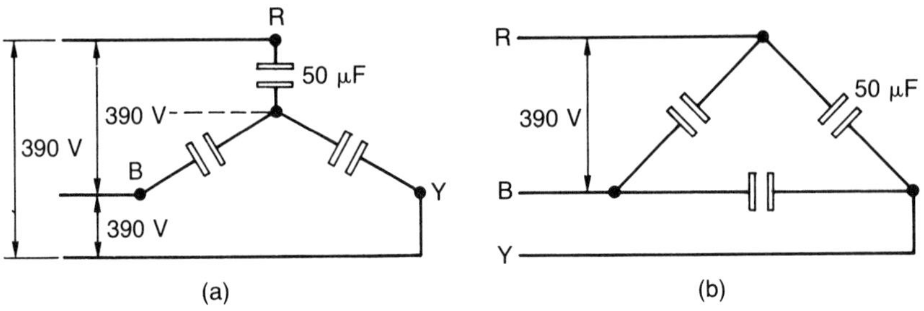

Figure 5.8 *Balanced three-phase (a) star-connected and (b) delta-connected loads for Example 5.5*

(a)   *Star-connection*

$Z_P = X_P$ alone; there is no resistance involved

$\therefore Z_P = 1/(2\pi f C) = 1/(2\pi \times 50 \times 50 \times 10^{-6}) = 63.66\,\Omega$

$V_P = V_L/\sqrt{3} = 390/\sqrt{3} = 225.2\,V$

$\therefore I_P = V_P/Z_P = 225.2/63.66 = 3.54\,A$

$\therefore I_L$ is also $3.54\,A$

(b)   *Delta-connection*

$Z_P = 63.66\,\Omega$ from part (a)

$V_P = V_L = 390\,V$

$I_P = V_P/Z_P = 390/63.66 = 6.13\,A$

$\therefore I_L = \sqrt{3} \times 6.13 = 10.62\,A$

Exercises

5.1   Three resistors, each of $40\,\Omega$, are connected in star to a $415\,V$, three-phase supply. Determine:
(a)   the phase voltage
(b)   the phase current
(c)   the line current.

5.2   Three resistors, each of $40\,\Omega$, are connected in delta to a $415\,V$, three-phase supply. Determine:
(a)   the phase voltage
(b)   the phase current
(c)   the line current.

5.3   Three $50\,\Omega$ resistors are connected to a three-phase supply (a) in star and then (b) in delta. For each of the two load configurations, calculate the total load power dissipation.

5.4   Three coils, each of inductance $150\,mH$ and resistance $50\,\Omega$, are connected in star to a $50\,Hz$, three-phase supply. If the line current is $4\,A$, calculate:
(a)   the phase voltage
(b)   the line voltage.

5.5   Three coils, each of inductance $150\,mH$ and resistance $50\,\Omega$, are connected in delta to a $400\,V$, $50\,Hz$, three-phase supply, Calculate:
(a)   the phase current
(b)   the line current
(c)   the total load power dissipation.

5.6 Each phase of a star-connected, balanced, three-phase load comprises an inductance of 159.2 mH and resistance 50 Ω. If the load is connected to a three-phase 400 V, 50 Hz supply, calculate the total load power dissipation and its power factor.

5.7 Three 47 µF capacitors, each in series with a 100 Ω resistor, are connected in delta across a 400 V, 50 Hz, three-phase supply. Calculate the power dissipated by the total load and the load phase angle.

5.8 A 415 V, 50 Hz, three-phase inductive load has a phase coil inductance of 0.2 H and resistance of 20 Ω. The load phase windings can be switched between star and delta connections. Calculate the current drawn from the supply in:
(a) the star connection
(b) the delta connection.

6 Measurements and measuring instruments

6.1 Introduction

The art of taking accurate electrical measurements and the instruments used are well documented in the many specialist textbooks which are readily available from several publishers. Therefore, the aim of this chapter is to give the mechanical engineer a brief insight into the use of simple instruments for the measurement of electrical current, voltage, resistance and frequency. Of the four quantities mentioned, perhaps the two most often measured are current (amps) and voltage (volts). The instruments. used for these tasks are called *ammeters* and *voltmeters* respectively. They both operate in much the same way by being electrical machines which either have a moving pointer or a row of changeable numbers (digits) that indicate the amount of current flowing through them.

6.2 Measurement of current and voltage

Figure 6.1(a) shows a simple electrical circuit in which current, I_1, and voltages, V_1 and V_2, are to be measured. Figure 6.2(b) shows how the ammeter and the voltmeters are connected to the circuit in completely different ways. The ammeter connection requires the circuit to be broken at the point of measurement and the ammeter to be inserted such that the full circuit current flows through it. On the other hand, the measurement of voltage using a voltmeter does not require the circuit to be broken nor does the voltmeter need the full circuit current to flow through it. On the contrary, the ideal voltmeter should take zero current by having an infinitely high internal resistance. The ideal ammeter, being in series with the circuit into which it has been inserted, should have a zero internal resistance. If the ammeter does have resistance, it will increase the total circuit resistance and consequently reduce the very current it is trying to measure.

Figure 6.1(b) demonstrates the method of connecting the two types of measuring instrument. The ammeter measuring I_1 is inserted in series with the circuit into a break made between points A and E. The voltmeters to measure voltages V_1 and V_2 require no circuit break and are connected externally across R_1 and R_2 between points A and B and between B and C respectively.

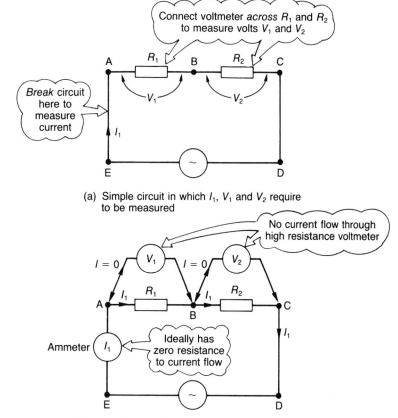

Figure 6.1 *Measurement of current and voltage*

(a) Simple circuit in which I_1, V_1 and V_2 require to be measured

The current (amps) to be measured must pass through the ammeter. The circuit must be broken between A and E and ammeter connected as shown.

Ideal voltmeters have an infinite internal resistance to prevent them from drawing current from the circuit under test.

(b) Measuring instruments connected

Hopefully, the resistances of the two voltmeters will be so high that they do not divert any of the original current flowing through R_1 and R_2. If the current flowing through the two resistors is reduced, by that diverted through the voltmeters, the voltages across the resistors, as read by the voltmeters, will be artificially low and an error reading will result.

6.2.1 Errors in current measurements

If the internal resistance of the ammeter is not truly zero, the insertion of the ammeter into the current-carrying circuit will introduce an additional resistance that will reduce the circuit current to be measured.

Example 6.1   *The circuit shown in Figure 6.2(a) is to have the current flowing measured using a moving coil ammeter. If the ammeter has an internal resistance of 5 Ω, estimate the error in the measurement.*

Before the measuring instrument, A, is connected in series with the circuit, the true current flowing is $12\,\text{V}/10\,\Omega = 1.2\,\text{A}$.

With the ammeter in the circuit, as shown by Figure 6.2(b), an additional $5\,\Omega$ meter resistance is introduced and the total circuit resistance becomes $(10 + 5)\,\Omega$. The current being measured by the ammeter is now only $12\,\text{V}/15\,\Omega = 0.8\,\text{A}$.

The error in the reading is the difference between the true and actual currents, that is, $(1.2 - 0.8) = 0.4\,\text{A}$. It is usual to express this error as a percentage of the true reading, so we have:

$$
\begin{aligned}
\text{Percentage error} &= (\text{error} \times 100)/(\text{true reading}) \\
&= 0.4 \times 100/1.2 \\
&= 33.3\%
\end{aligned}
$$

Clearly, this size of error in any measurement is unacceptable and the conclusion is that *the ammeter used to measure current must have an insignificant resistance compared with that of the circuit into which it is being inserted.* For example, had the circuit resistor in Figure 6.2(a) been $1000\,\Omega$ rather than $10\,\Omega$, the error would have been only 0.5%.

6.2.2 Errors in voltage measurements

Unlike the current measuring ammeter, *the voltmeter must have an internal resistance which is much higher than that of the circuit across which it is attempting to measure potential difference* in volts.

Example 6.2   *A moving coil voltmeter of 5 Ω internal resistance is to be used to measure the voltage drop across the 20 Ω resistor of Figure 6.3(a). Estimate the percentage error in the voltmeter reading.*

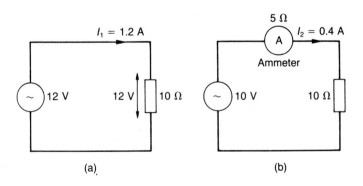

Figure 6.2 *Error in current measurement caused by measuring instrument*

(a)   (b)

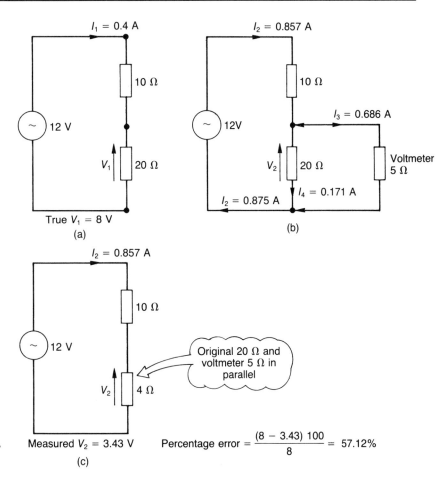

Figure 6.3 *Error in voltage measurement caused by relatively low voltmeter resistance*

The measurement circuit looks like Figure 6.3(b), which can be simplified to Figure 6.3(c).

The true voltage across the $20\,\Omega$ resistor is:

$$12\,\text{V} \times 20\,\Omega/(20+10)\,\Omega = 8\,\text{V}$$

With the $5\,\Omega$ voltmeter connected across the $20\,\Omega$ resistor, the latter is effectively reduced to $4\,\Omega$. The measured voltage, V_2, is developed across this $4\,\Omega$ resistance as:

$$12\,\text{V} \times 4\,\Omega/14\,\Omega = 3.43\,\text{V}$$

The percentage error in the actual reading is:

$$\frac{100 \times \text{error}}{\text{true}} = \frac{100 \times (8-3.43)}{8} = 57.12\%$$

Had the voltmeter internal resistance been much higher than that of the $20\,\Omega$ resistor, say $1000\,\Omega$, a repeat of the above calculations would show the measurement error to be only 0.625%.

Figures 6.3(a) and (b) illustrate the unsettling effect of shunting (connecting in parallel) the $20\,\Omega$ resistor with a low voltmeter resistance of only $5\,\Omega$. Because the effective overall circuit resistance is reduced from $30\,\Omega$ to $14\,\Omega$, the true circuit current of 0.4 A increases to 0.875 A. However, the current passing through the $20\,\Omega$ resistor, which generates the voltage to be measured, has nevertheless been reduced from 0.4 A to 0.171 A. The reason is that the 0.875 A current finds the $5\,\Omega$ voltmeter resistance an easier path to follow than the $20\,\Omega$ resistor and divides accordingly: 0.686 A through the voltmeter and only 0.171 A through the $20\,\Omega$ resistor.

6.3 The moving coil or analogue multimeter

Figure 2.8 in Chapter 2 shows how a loop conductor, if mechanically rotated in a permanent magnetic field flux, generates a current which flows around the loop. The device is known as a *generator*. However, if the loop of the same generator is not made to rotate from an external source but instead is supplied with a d.c. current, the loop experiences a turning force. This is caused by the interactive forces set up between the permanent magnetic flux and the flux produced by the current in the loop. The device is now acting as a d.c. *motor*. The turning force, or torque, exerted on the loop conductor is directly proportional to the amount of current supplied to the loop. If the movement of the loop were made to compress a coiled spring, the greater the current in the loop the further the loop would turn before the compressed spring exerted sufficient force to stop it.

The fitting of a pointer to the rotating loop, to indicate the stopped position against a graduated scale, would enable the device to be used as a *moving coil meter*. The scale could be graduated in amps to indicate the amount of current in the loop or coil and the device would be an *ammeter*.

On the other hand, since the resistance of the coil will be known, the voltage required to drive a particular current through the coil can be calculated. Thus, the scale can be graduated in volts and the device has become a *voltmeter*.

Finally, if the case containing the moving coil mechanism is fitted with an internal battery, this battery can be made to drive current through the moving coil and an external unkown resistance and the device has become an ohmmeter for measuring the value of the external resistance. The instrument is truly multi-purpose and is known as a *multimeter*.

Figure 6.4 shows the popular AVO Model 8 analogue multimeter which can be found in most electrical and electronic workplaces. The multimeter can be switched to measure either d.c. or a.c. quantities. The meter movement, being a d.c. motor, does not respond correctly to alternating currents. Any a.c. fed into the meter is first passed through

Figure 6.4 *Analogue multimeter*

an internal full bridge diode rectifier before being measured by the moving coil assembly as direct current.

6.3.1 The ammeter

Figure 6.5 is a diagrammatic representation of the moving coil meter configured to act as an ammeter. Any d.c. to be measured is fed into the meter through the positively marked terminal. If the current is too large for the moving coil to handle without overheating, only a known

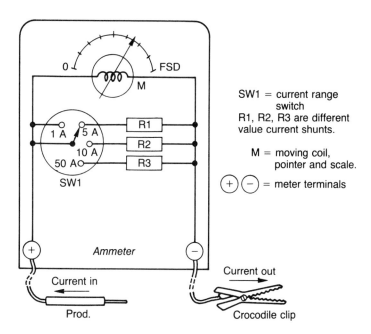

Figure 6.5 *Moving coil meter multimeter switched to ammeter role*

safe fraction of the total current is passed through the coil. The remainder of the current is shunted around the coil out of harm's way. Switch SW1 and resistors R1, R2 and R3 are the current range switch and shunting resistors.

Example 6.3 *A 5 Ω meter movement has a full scale deflection (FSD) when passing 1 A. It is required to be used as an ammeter with the FSD of the pointer indicating a current of 5 A. Calculate the value of the internal coil shunt resistor for this situation.*

This problem applies to the ammeter shown in Figure 6.5. With the current range switch SW1 in the 5 A position, we need to calculate the value of R1 so that it takes all of the current which exceeds the coil safe current value of 1 A. Thus the shunt must take a current of $5A - 1A = 4\,A$.

In order to calculate the value of R1, we now need to know the voltage which will appear across it. This, of course, is the same as that across the meter coil, namely $1\,A \times 5\,\Omega = 5\,V$.

The value of R1 is the current it must pass divided by the voltage across it:

$R1 = 5\,V/4\,A = 1.25\,\Omega$

Similar calculations for the 10 A and 50 A current ranges of SW1 give values for R2 and R3 of $0.5555\,\Omega$ and $0.102\,\Omega$ respectively.

6.3.2 The voltmeter

Figure 6.6 shows the internal configuration of the multimeter in the voltmeter role. Assuming the moving coil assembly to be the same one as used in Example 6.3, the meter will show FSD for an applied voltage of only 5 V. The range of the meter is extended by using switch SW2 to connect one of the resistors R4, R5 or R6 in series with the 5 Ω coil. The inserted resistor removes, or 'drops', the excess voltage over 5 V which would otherwise cause damage to the moving coil assembly.

Example 6.4 *Calculate the value of resistor R5 required to extend the range of the previous 5 Ω meter movement to measure 100 V FSD.*

The basic meter will take only 1 A at 5 V for FSD. Any voltage above 5 V must be dissipated elsewhere before being applied to the moving coil assembly. Therefore, in order to measure 100 V, 95 V must be dropped across the inserted series resistor, R5, leaving only 5 V for the moving coil to handle safely.

The current which passes through the coil at FSD has been given as 1 A and since this same current passes through R5 we can calculate the value of R5 to give the required 95 V drop across it:

$R5 = 95\,V/1\,A = 95\,\Omega$

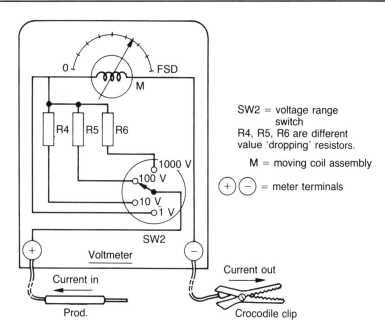

Figure 6.6 *Moving coil multimeter switched to voltmeter role*

Similar calculations can be made to determine the values of R4 and R6 for the 10 V and 1000 V ranges. These will be found to be 5 Ω and 995 Ω respectively.

6.3.3 The ohmmeter

The internal multimeter arrangement is shown in Figure 6.7. Because the method of measuring resistance entails sending a current through the unknown resistance, the multimeter contains a battery, B, which is switched into circuit for the purpose. The principle of operation is that the known internal battery voltage pushes an amount of current through the unknown resistance, connected externally between the meter terminals, which depends upon its value. Therefore, it is a simple matter to calibrate the meter, which responds to current flow, directly in ohms.

Since the battery voltage will fall with use and age, an extra internal adjustable resistance, Z, is used to compensate for this. Before each measurement the crocodile clip and the prod must be shorted together and the heavy current that flows is regarded as the zero resistance current. A special calibration mark on the meter scale (near FSD) indicates the pointer position. Z is adjusted by the *set-zero* control, on the multimeter casing, so that the pointer is exactly on the zero resistance mark. The unknown resistance is then connected between the crocodile clip and the prod and the new pointer position indicates the resistance value on the calibrated scale.

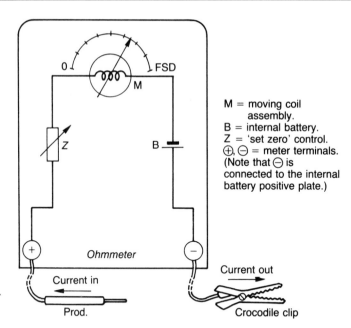

Figure 6.7 *Moving coil multimeter switched to ohmmeter role*

M = moving coil assembly.
B = internal battery.
Z = 'set zero' control.
⊕, ⊖ = meter terminals.
(Note that ⊖ is connected to the internal battery positive plate.)

It should be noted that *the multimeter supplies current from its negative terminal.* This should be taken into account when using the multimeter to check polarity-sensitive semiconductor diodes and transistors.

6.4 The digital multimeter

Many users find the modern digital multimeter, such as the one shown in Figure 6.8, easier to use than the analogue version described in the previous section. One reason is that there is no pointer position to interpret against the correct scale from a multiplicity of scales. In the place of a pointer, the digital meter displays a simple row of numbers, or digits, which do not suffer from errors caused by parallax and indicate the quantity being measured directly.

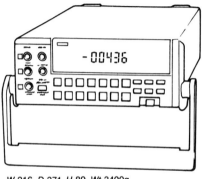

Figure 6.8 *Digital multimeter*

W 216 D 371 H 89 Wt 3400g

Unlike the moving coil meter, the digital meter does not require a current input mechanically to move a pointer assembly. Therefore it takes virtually no current from the circuit being measured, giving it an improved accuracy of measurement.

6.5 The signal generator

Figure 6.9 shows a typical signal generator. It is no more than an alternator which runs at a very high speed and produces sine wave voltages and currents used for testing electronic equipment. If the signal generator will additionally produce other than sine wave outputs, such as square waves or triangular waveforms, it is called a *function generator*.

A typical use of a signal generator is as a signal source to feed pure sinusoidal voltages into an electronic amplifier. The amplifier output is at the same time monitored using an oscilloscope. The oscilloscope trace will show how well the amplifier magnifies the input signal and whether it is producing distortion in the process. Precisely how the oscilloscope works is described in the next section.

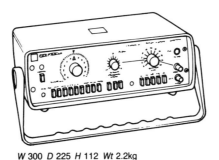

Figure 6.9 *Signal function generator*

W 300 D 225 H 112 Wt 2.2kg

6.6 The oscilloscope

6.6.1 Construction

A typical oscilloscope front panel arrangement is shown in Figure 6.10. The essentials of the oscilloscope are a cathode ray tube (CRT), an adjustable gain amplifier and a variable speed timebase generator. Figure 6.11 shows a diagrammatic representation of how these components interrelate.

The CRT is an evacuated glass tube. It is flared at one end to form the screen which has a fluorescent phosphor coating on its inside. The other end of the CRT holds what is known as an *electron gun*. It fires electrons as 'bullets'. This gun comprises a heater coil, H, which makes a tungsten plate, called the cathode, K, red-hot. The heated cathode, being in a vacuum, emits a cloud of electrons which congregate as a negative *space charge* near the surface of the cathode. This process is known as *thermionic emission*. An anode, A, having a large positive d.c. voltage

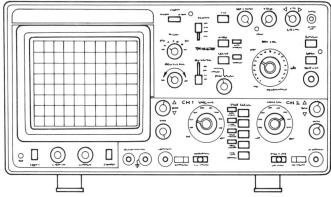

Figure 6.10 *Dual trace oscilloscope*

W 310 D 400 H 160 Wt 9kg

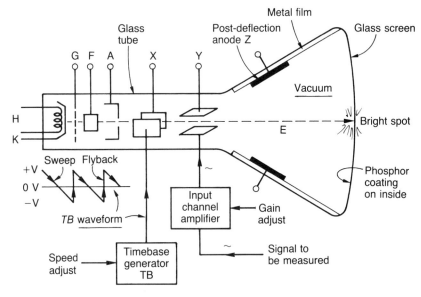

Figure 6.11 *Principle of cathode ray tube*

H   heater coil
K   cathode (emits electron beam)
G   grid (brightness)
F   focus anode (sharp trace)
A   main anode (accelerates electrons)
X   plates for horizontal beam deflection (sweep + flyback)
Y   plates for vertical beam deflection (signal input)
E   electron beam (produces trace on screen)

(typically 2000 V), attracts the negatively charged electrons from the space charge. The electrons rush to the anode, passing through a focusing tube, F, on their way and arrive at the anode which has a hole drilled through it at the expected point of impact. The speeding electrons pass through the anode hole and are 'fired' along the axis of the CRT until they strike the phosphor-coated screen. The impact of the electron stream causes the phosphor to glow, producing a bright spot of light which can be seen from the outside of the CRT.

The stream of negative electrons, E, passes between two pairs of deflecting plates. If one plate of a pair is made positively charged and the other negatively charged, the electrons deflect from their straight path towards the screen and strike the screen off-centre. A negative deflecting plate pushes the electron beam away from it; a positive plate attracts the beam. The greater the deflecting voltage applied to the plates, the greater the movement of the point of light away from the screen centre. In practice, the outside of the screen has a transparent plastic overlay of 1 cm squares called a *graticule*. This enables the displacement of the light spot to be measured and calibrated to indicate the size of the voltage applied to the deflecting plates. The plates which cause the electron beam to deflect up and down are called the *Y-deflection plates*. The other pair of deflecting plates deflect the beam left or right of screen centre and are called the *X-deflection plates*. The faster the beam of electrons strikes the screen, the brighter the trace caused. A post-deflection accelerator anode, Z, is often fitted in the CRT to produce a brighter trace. It is fitted after the beam deflection has occurred because the slower the electron speed the easier it is to cause it to deflect. The brightness of the screen trace can be varied by adjustment of a small negative voltage applied to the grid electrode, G, located close to the cathode in the electron gun. It controls the density of the emitted electron beam and we shall see later how it is used to stop the beam completely during the trace flyback period.

6.6.2 Operation

The function of the *timebase generator* is to supply a voltage having a waveform as shown in Figure 6.12. This sawtooth waveform is applied to the X-deflection plates and gives the electron beam a repetitive linear sweep and flyback action. At point A on the waveform, the horizontal

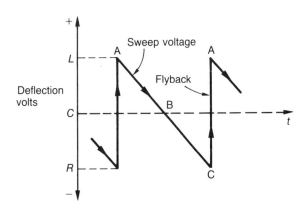

Figure 6.12 *Timebase generator output waveform fed to* X-*deflection plates*

deflection plates attract the electron beam to the extreme left of the screen. At point B, the beam is at the screen centre, while at point C the beam has been pushed to the extreme right. When the spot reaches the extreme right, a large negative voltage is applied to the grid. This cuts off the electron beam, and the screen trace is extinguished while the beam is made to fly back very rapidly to the screen left. The grid blanking pulse is then removed and the trace reappears ready to start another sweep cycle. The time which the trace takes to be deflected across the screen is adjustable by a timebase control knob located on the front panel. The timebase control is calibrated in sec/cm. A setting of $5\,\mu s$, for example, means that the trace takes $5\,\mu s$ to sweep 1 cm horizontally.

The Y-deflection plates are fed with the signal voltage which is to be displayed on the screen. The signal may be too small to cause an appreciable deflection on the screen Y-axis so an adjustable gain *Y-channel input amplifier* is provided to amplify the weak signal. The amplifier has a front panel gain control knob calibrated in volts/cm. A setting to 2 V, for example, means that an input signal of 2 V d.c. will cause the trace to move vertically by 1 cm. Usually an upward movement indicates a positive input signal, a downward movement a negative signal.

If a varying signal voltage is applied to the Y-plates while the X-plates are causing the beam to deflect linearly across the CRT screen face, the result is a trace displaying the instantaneous signal amplitude with time. Figure 6.13 shows a selection of the different displays one can obtain on the oscilloscope screen. Assuming that the front panel controls are set, for example, as follows:

X-plate deflection timebase $= 5\,ms/cm$
Y-plate signal input amplifier $= 2\,V/cm$

the various waveform traces represent input signal voltages as indicated below:

(a) A zero signal input
(b) $2\,V/cm \times 3\,cm = +6\,V$ (d.c.)
(c) $2\,V/cm \times -3\,cm = -6\,V$ (d.c.)
(d) $A = 2\,V/cm \times 3\,cm = 6\,V$
     $T = 5\,ms/cm \times 8\,cm = 8\,cm = 40\,ms$
     $f = 1/T = 1/40\,ms = 25\,Hz$

   This is a sinusoidal voltage given by:

   $$V = 6\sin 50\,\pi t \text{ V}$$

(e) This is a sine wave of the same amplitude but twice the frequency of (d) so the trace represents:

   $$V = 6\sin 100\pi t \text{ V}$$

(f) This is a square wave of amplitude 6 V and frequency 33.33 Hz.

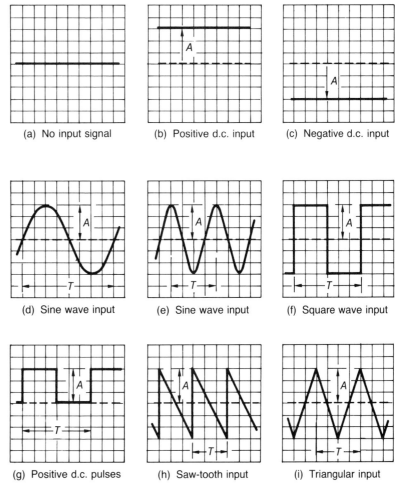

(a) No input signal (b) Positive d.c. input (c) Negative d.c. input

(d) Sine wave input (e) Sine wave input (f) Square wave input

(g) Positive d.c. pulses (h) Saw-tooth input (i) Triangular input

A = waveform amplitude; $\frac{1}{T}$ = waveform frequency

The graticule overlaying the trace represents 1 cm squares

Figure 6.13 *Oscilloscope traces for different input waveforms*

(g) These are positive 6 V d.c. pulses at a pulse repetition frequency (p.r.f.) of 33.33 pulses per second (p.p.s.).

(h) This is a sawtooth waveform with a peak-to-peak value of 12 V and a sweep time of $3\,cm \times 5\,ms/cm = 15\,ms$.

(i) This is a triangular wave of amplitude 6 V and frequency 50 Hz.

Exercises

6.1 A 47 kΩ resistor and a 220 kΩ resistor are connected in series across a 9 V battery. If the multimeter used to measure the voltage developed across the 220 kΩ resistor is:

    (a)   a moving coil meter having an internal resistance of 200 kΩ

    (b)   a digital meter having an internal resistance of 1 MΩ

    Calculate the percentage error in each case.

6.2   You have a small moving coil multimeter which has a coil resistance of 100 Ω and gives a FSD when passing a current of 100 mA. Calculate:

    (a)   the value of the shunt resistor needed to enable the meter to measure a current of 5 A

    (b)   the value of the series resistor needed to convert the meter to a 100 V voltmeter.

6.3   Refer to Figure 6.13. For the screen traces shown, calculate the amplitude and frequency of the waveform if the X-plate timebase and Y-plate input amplifier controls were set as follows:

For (a), (d), (g): $X = 2\,\text{ms/cm}$, $Y = 5\,\text{V/cm}$

      (b), (e), (h): $X = 5\,\mu\text{s/cm}$, $Y = 5\,\text{mV/cm}$

      (c), (f), (i): $X = 20\,\text{ms/cm}$, $Y = 2\,\text{V/cm}$

7 Semiconductor diodes

7.1 Semiconductor material

Section 1.1.4 explained how the electrical conductivity of a material depends upon the nature of its atomic structure. If the number of electrons orbiting in the outer energy shell of the atoms is well short of the number required to fill that shell, the material is classed as a good electrical conductor. This is because the odd electron in a near empty outer shell is easily detached from its parent atom by the application of an external voltage, heat or even light. The detached electron is classed as a *free electron* and wanders aimlessly around the material. However, if the material is subjected to a potential difference, any free electrons, being negative charges, drift towards the more positive potential and so form an electric current. If the outer shell of the atom is completely filled, electrons are very difficult to dislodge and the material acts as a good insulator.

7.1.1 Pure silicon semiconductor

Silicon is a material which has an outer shell with a capacity for eight orbiting electrons but actually contains only four. The outer shell in fact *appears* to be filled because each of the four electrons belonging to the parent silicon atom orbits not only its own atomic nucleus, but also that of an immediately adjacent atom. These shared electrons form *valency bonds* between each silicon atom and its four closest neighbours. By this means a diamond-shaped crystal lattice of silicon is formed. At normal room temperatures, the silicon material is found to be neither a good insulator nor a good conductor of electricity; it is a *semiconductor*.

Figure 7.1 shows how a model of the pure silicon lattice can be constructed using ping-pong balls and fish-hooks. The balls represent the silicon atom nucleus together with its inner shell electrons having a net electrical charge of $+4$ which is balanced by the -4 of the four hooks, each of which represents a valency electron. Thus, the silicon atom is electrically neutral. The model is completed by linking the hooks, so joining the balls together. The four linked hooks, of course, represent the valency bonds holding the silicon lattice in place. The pure silicon atom is said to be *tetravalent*.

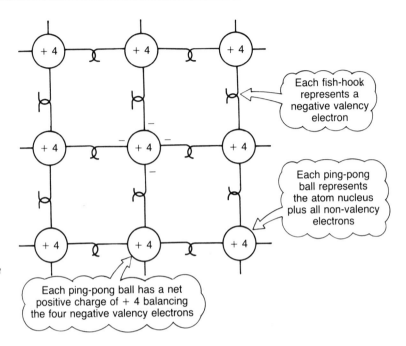

Figure 7.1 *Model of pure silicon atomic lattice made from ping-pong balls each fitted with four fish-hooks*

The valency bonds between atoms are not easy to break. However, if the temperature of the silicon is raised sufficiently, some of the orbiting valency electrons gain enough speed to break free. Now, any free electron must leave its parent atom short of one electron. The parent atom effectively has a positive *hole* one electron charge in magnitude. The free electrons wandering about the silicon find holes in other atoms and combine with them. At the same time other free hole–electron pairs are being generated. The higher the silicon temperature, the greater the number of free electrons and holes.

The thermal generation of hole–electron pairs can be simulated by imagining a large, horizontal table, the surface of which is covered by shallow depressions (holes), and into which ping-pong balls are placed. If the table is mechanically vibrated sufficiently violently, some of the balls will leave their holes and wander around the table. Some of the wandering free balls will come across empty holes and fall into them. At the same time more balls will be being freed. If the table is now lifted at one end to simulate a negative potential being applied to that end and a positive potential to the other end, the free balls will roll down the table and the holes will appear to move up. The moving balls and holes together constitute a simulation of the *intrinsic current* flow made possible in pure silicon by a sufficiently high temperature producing the necessary *thermal agitation*.

A further simulation of how the fixed holes can be regarded as being mobile current carriers is shown in Figure 7.2. The situation is that in the waiting room at a doctor's surgery. In (a), there are six patients occupying six chairs. In (b), the doctor calls for the next patient, the patient goes

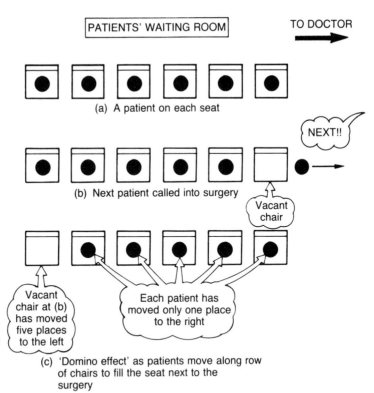

Figure 7.2 *Doctor's waiting room model of movement of holes and electrons in a semiconductor. The patients represent electrons and the chairs represent holes*

and the right end chair becomes vacant. The patients then all move one chair to the right so that, in (c), it is now the left end chair that is vacant. The patients (electrons) have each made a small movement and this has resulted in a vacant chair (hole) making a large movement.

7.1.2 Doped silicon semiconductors

The pure silicon semiconductor is not much use at room temperatures. It acts more like an insulator than it does a conductor. However, it has been found that the addition of a minute quantity of impurity to the pure silicon dramatically increases its low-temperature conductivity. The effect of the added impurity is to produce more current carriers at normal room temperature. One type of impurity donates more mobile electron carriers to the pure silicon to produce what is known as n-type silicon. Another class of impurity adds mobile holes making p-type silicon. The addition of the impurity is called *doping*. It should be noted that the adding of more mobile electrons or holes is done in such a manner so as not to upset the electrical neutrality of the pure silicon. The explanation of how this is achieved now follows.

n-type silicon semiconductor

The model for this semiconductor material is shown in Figure 7.3. One in a hundred million tetravalent silicon atoms is replaced by a pentavalent atom such as that of phosphorus, arsenic or antimony. The pentavalent dope atom has a positively charged centre of +5 which is electrically balanced by five fish-hooks or valency electrons. However, one of the five hooks cannot find a spare tetravalent hook with which to link. So the surplus tetravalent hook, or electron, is not firmly bonded to the system and is very readily made to break away and wander off on its own. However, once the spare valency electron has left its parent dope atom, the latter becomes positively charged because it now has only four electrons to try and balance its +5 centre. The dope atom effectively has become a *fixed* positive charge, or hole, into which no electron can readily fall and which therefore plays no further part in forming a moving electric current. On the other hand, the wandering free electron, donated to the system by the pentavalent dope, is available to act as a current carrier. It readily travels through the silicon structure towards any positive potential which may be provided. Even though the n-type material has a surplus of mobile electrons as current carriers, it is still overall electrically neutral. There are as many fixed positive charges as there are mobile electrons.

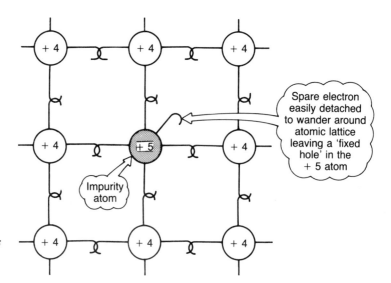

Figure 7.3 *Model of pentavalent impurity atom donating a surplus electron to the pure tetravalent silicon lattice*

p-type silicon semiconductor

Figure 7.4 shows the atomic lattice where a trivalent impurity, such as aluminium, gallium or indium, is used to dope the pure silicon. The +3 atom centre and its three valency electrons are once again misfits in the tetravalent structure. Figure 7.4(a) depicts the hole left by the 'missing'

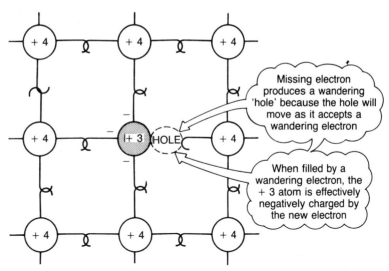

(a) Trivalent impurity atom produces surplus mobile hole

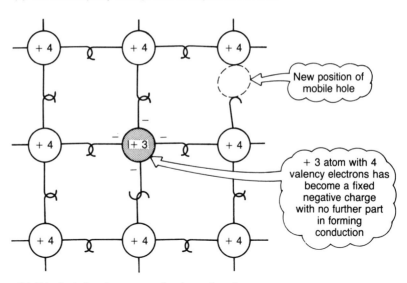

Figure 7.4 *Model of trivalent impurity atom accepting an electron from the pure tetravalent silicon lattice to produce a surplus mobile hole*

(b) Trivalent atom becomes a fixed negative charge

electron associated with the dope atom. When a wandering free electron, caused by thermal agitation, passes by, it falls into the hole next to the +3 atom. The additional electron now links with the vacant hook of the adjacent silicon atom and at the same time turns the +3 dope atom into a *fixed* negative charge because it now has four electrons. The surplus hole originally introduced by the dope atom has effectively moved to whichever silicon atom donated the free electron that filled it. This effect is shown in Figure 7.4(b). Once again, the p-type doped silicon, even with its surplus positive holes as mobile current carriers, is overall electrically neutral. There are as many fixed negative charges as there are free positive holes.

7.2 Operation of the p–n junction

7.2.1 Formation of the p–n junction

Figure 7.5 summarizes the situation with n-type and p-type silicon. The n-type has electrons as the current carrier and the p-type has holes as the current carrier. Both the n-type and p-type materials are electrically neutral.

Figures 7.6(a) and (b) show the effect of forming a junction by bringing together a piece of n-type and a piece of p-type to form a p–n junction. Once the electrically neutral pieces in (a) touch each other in (b), the freely wandering current carriers in each piece come to the p–n boundary and simply pass across it. Since both n-type and p-type are electrically neutral there is no reason for these initial boundary crossings not to happen. However, once the negative electrons from the n-type arrive in the neutral p-type, the p-type becomes negatively charged. Eventually the p-type becomes so negatively charged that it repels any further electron migration from the n-type. A similar state of affairs exists in the n-type material caused by the migration of positive holes.

The overall result is that a voltage barrier between the n-type and the p-type is established and no further movement of current carriers across the boundary is permitted. The voltage is found to be 0.6 V, with the n-type being the more positive. Further, there are no mobile current carriers to be found in the immediate vicinity of the p–n boundary. The area depleted of its normal current carriers is called the *depletion*

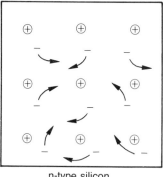

n-type silicon

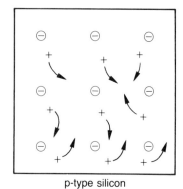

p-type silicon

Figure 7.5 *Representation of n-type and p-type doped silicon for improved conduction*

$\oplus$ = fixed pentavalent impurity atom
$\ominus$ = fixed trivalent impurity atom
$+\!\nearrow$ = mobile hole current carrier
$-\!\searrow$ = mobile electron current carrier

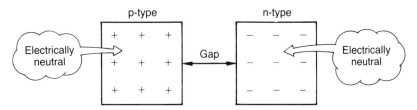

(a) A piece of p-type and a piece of n-type showing only the mobile current carriers in each. Balancing fixed charges are not shown.

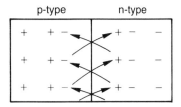

(b) Electrons wander from the neutral n-type into the neutral p-type making it negatively charged. The n-type becomes positively charged because of migrating holes from the p-type.

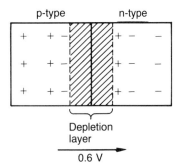

(c) A barrier potential of 0.6 V is established across the depletion layer which contains no mobile current carriers.

Figure 7.6 *The p–n junction*

layer. It acts as if it were a large resistance to current flow. That, of course, is exactly what it is, since it now has no means of carrying current. Figure 7.6(c) illustrates the situation.

7.2.2 Biasing the p–n junction

Figure 7.7(a) illustrates the unbiased (no external voltage applied) junction with its usual depletion layer. In Figure 7.7(b), a positive potential

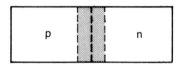

(a) Unbiased p-n junction − normal depletion layer.

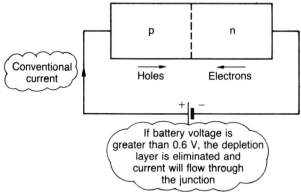

(b) Forward biased p–n junction – the eliminated depletion layer
    allows current to flow

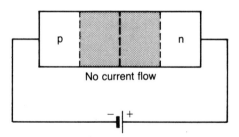

Figure 7.7 *Biased p–n junction
acts as an electronic one-way
valve. It is called a diode*

(c) Reverse biased p–n junction – increases the width of the
    depletion layer even further preventing current flow

has been applied to the p-type and a negative to the n-type. This is
called *forward biasing* the junction. If the applied voltage is less than
0.6 V, there is no flow of current through the material. If the applied
voltage is made equal to 0.6 V a trickle of current across the junction is
observed. Further increase of the applied bias voltage causes the current
flowing to increase very rapidly indeed. With the applied voltage at only
0.8 V to 0.9 V the junction conducts almost as if it had zero resistance.
What has happened is that the first 0.6 V of the applied voltage has
reduced the barrier potential and the depletion layer width to zero so
allowing conventional current to flow across the junction from p-type to
n-type.

With the battery connected the other way around, as shown in Figure 7.7(c), the junction is said to be *reverse biased*. The effect of the negative battery voltage on the p-type, and battery positive on the n-type, is for the effect of the depletion layer to be enhanced. The depletion layer is widened and conduction through the reverse biased junction is made even less possible.

The p–n junction is given a special name. It is called a *semiconductor diode*. It has the important property of acting as a 'one-way street' for the flow of current through it. Current will flow only from the p-type to the n-type and, even then, only provided the applied voltage is greater than 0.6 V.

7.3 The semiconductor diode

Figure 7.8(a) shows the circuit symbols used for the diode. Note how the arrow-shaped body points in the direction of conventional current flow through it. The positive electrode, or terminal, of the diode is called the *anode* and the other terminal is called the *cathode*. For the diode to conduct, the anode must be made more positive than the cathode and, even then, by more than 0.6 V. Figure 7.8(b) shows the diode characteristic curve of current flow against applied voltage. Note how the current in the forward direction does not start to flow until the 0.6 V are applied and then how rapidly the current increases thereafter. In the reverse direction, there is virtually no conduction until the voltage has reached such a large value that it breaks down the diode structure and a destructive reverse current flows.

7.3.1 The diode as a half-wave rectifier

The one-way effect of the diode is put to use in the rectification, or changing, of alternating currents to direct currents, that is, a.c. to d.c. If an alternating voltage is applied to the anode of a diode, as shown in Figure 7.9, it will be only during the positive half-cycles that the diode is correctly biased for conduction. It will be only during the positive half-cycles of the applied voltage that the anode of the diode will be more positive than the cathode, so allowing current to flow from the anode to cathode. The negative half-cycles of the alternating input voltage will effectively reverse bias the diode and it will not conduct. The negative half-cycles will be removed and will not appear at the diode output. Therefore, the output voltage waveform at the diode cathode will be a series of d.c. pulses each the shape of the input positive half-cycle. This process of producing d.c. from a.c. is called *half-wave rectification*.

Figure 7.9(b) shows a method of using a reservoir, or smoothing capacitor, *C*, to even out the rough d.c. produced by the half-wave rectification process. The principle involved is that the first positive half-wave to pass through the diode sends a pulse of current to both the load resistor and the capacitor. The current fed into the capacitor develops a voltage

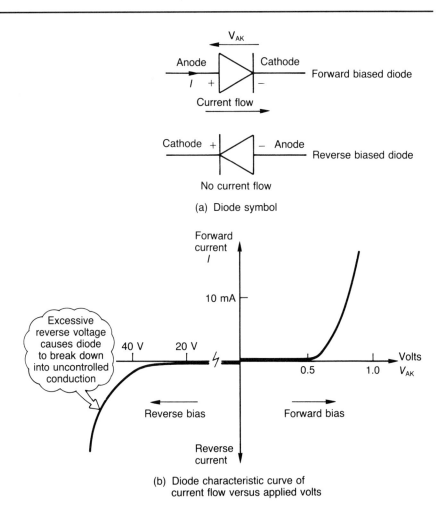

Figure 7.8

(a) Diode symbol

V_{AK}

Anode | Cathode — Forward biased diode

I + −

Current flow

Cathode + − Anode — Reverse biased diode

No current flow

Forward
current
I

10 mA

Excessive
reverse voltage
causes diode
to break down
into uncontrolled
conduction

40 V   20 V

0.5   1.0   Volts V_{AK}

Reverse bias   Forward bias

Reverse
current

(b) Diode characteristic curve of
current flow versus applied volts

across its plates which is retained after the current pulse from the diode
has subsided. The voltage retained by the capacitor provides a continued
supply of current to the load in the absence of an output voltage from the
diode during the negative half-cycle of the input waveform. The voltage
of the capacitor falls somewhat before the next positive pulse from the
diode arrives to replenish it. The amount the capacitor falls between
pulses is dependent upon the amount of current drawn from it by the
load. Generally, the capacitor holds the output d.c. at a designed average
level with the inter-pulse capacitor voltage variations causing a superimposed *ripple voltage*.

7.3.2 Diode bridge as a full-wave rectifier

The work of the smoothing capacitor can be made easier if the rectifier
produces d.c. pulses for both the positive and the negative half-cycles of

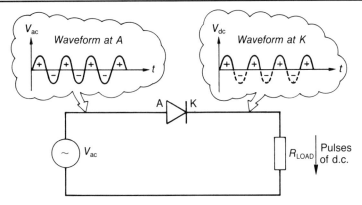

Diode rectifier cuts-off the negative half-cycles of the input a.c. waveform to produce rough d.c. in pulsed form

(a) Basic half-wave rectifier circuit

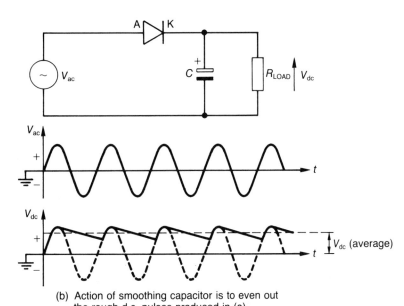

Figure 7.9 *Single diode used for half-wave rectification*

(b) Action of smoothing capacitor is to even out the rough d.c. pulses produced in (a)

the a.c. input voltage. This is achieved using the circuit shown in Figure 7.10(a). The diode bridge is fed by the input a.c. waveform shown in Figure 7.10(b).

During the positive half-cycles of the input waveform, point A supplies current to the diode bridge, the direction of current flow being shown by the solid arrows. The current passes in the forward direction through diode D1, from C to D, through the load resistor, and returns to point B via D2. During this time, the two diodes D3 and D4 are effectively reverse biased and do not conduct.

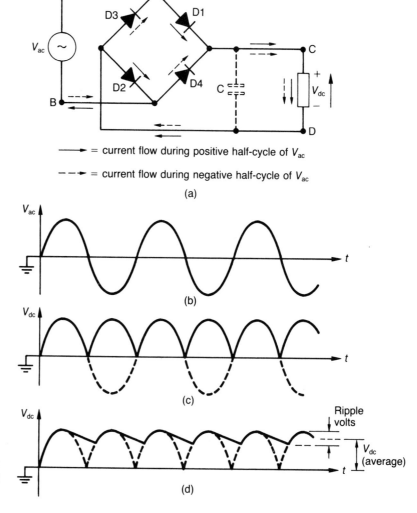

= current flow during positive half-cycle of V_{ac}

= current flow during negative half-cycle of V_{ac}

(a)

(b)

(c)

(d)

Figure 7.10 *Full-wave rectification using a diode bridge circuit. (a) The circuit; (b) the a.c. voltage input waveform; (c) the d.c. voltage output without C fitted; (d) the d.c. voltage output with C fitted.*

During the negative half-cycles of the input, point B supplies current to the diode bridge and the direction of current flow is shown by the dotted arrows. Note how the current again passes through the load resistor in the direction from C to D. Only diodes D3 and D4 conduct during the negative half-cycle.

Figure 7.10(c) shows the effect of full-wave rectification. The negative half-cycles of the input are now put to good use. The period between the positive pulses fed to the smoothing capacitor is less than that in the half-wave case and so the ripple voltage is correspondingly less and the average d.c. produced is greater.

8 Transistors

8.1 Introduction

In Chapter 7, it was explained that if a silicon p–n junction diode has a voltage of more than 0.6 V applied to it in the *forward* direction, it will pass current, but if the applied voltage is *reversed*, the junction acts as a complete block to current flow. We shall now see how a bipolar junction transistor (BJT) can be made using pieces of p-type and n-type silicon to form a pair of p–n junctions which together operate in a special way. Figures 8.1(a) and (b) respectively show the p–n–p and n–p–n transistors. In both cases, the three pieces of doped silicon are called the *emitter*, the *base* and the *collector*. The emitter-to-base junction is forward biased for conduction to occur, while the collector-to-base junction is heavily reverse biased, but nevertheless readily passes current. The battery connections show how the necessary junction biasing is achieved and the direction of the flow of conventional current is indicated by the arrows, I_E, I_B and I_C.

Applying Kirchhoff's first law to Figure 8.1, we can see that the current, I_E flowing in the emitter region is the sum of the currents, I_B and I_C, flowing in the base and collector regions.

$$I_E = I_B + I_C \qquad\qquad [8.1]$$

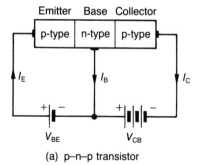

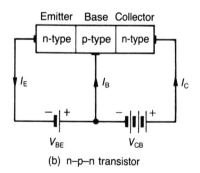

Figure 8.1 *Three different regions of p-type and n-type can form a bipolar transistor of p–n junctions. For conduction, one junction (BE) is forward biased, the other (CB) is reverse biased*

(a) p–n–p transistor

(b) n–p–n transistor

8.2 Operation of the bipolar junction transistor (BJT)

Figure 8.2 shows a more detailed sketch of the p–n–p bipolar junction transistor, or just transistor for short. The reason the term bipolar is included in its full name is because it describes the way the device

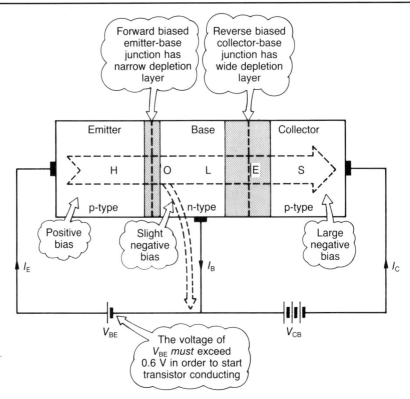

Figure 8.2 *Operation of the p–n– p silicon transistor*

conducts. The p-type silicon has a surplus of free positive holes and so, in the p-type, holes are the majority current carriers – conduction is by the flow of holes. In the n-type, free negative electrons are the majority current carriers and so conduction is mainly by the flow of electrons. Overall, then, the conduction of current through the p–n–p transistor depends upon both positive and negative majority carriers; the carriers are *bipolar*.

In the p–n–p transistor, the emitter and collector are doped to produce surplus mobile holes but the base region between is doped to produce a surplus of mobile electrons. Also, the thickness of the base region is not as shown in Figure 8.2. It is much thinner than either the emitter or the collector regions and it is this relative thinness which is an important factor in the operation of the transistor which is described next.

The battery, V_{BE}, forward biases the emitter–base junction to such a degree (0.6 V) that the junction depletion layer is reduced to zero width. The battery positive potential on the emitter repels the surplus positive holes in the emitter region towards the base region. The base region, being negatively biased by V_{BE} with respect to the emitter, attracts the positive holes repelled by the emitter. With the junction depletion layer having been effectively removed, the holes rush from the emitter into the narrow base region. Now the base–collector junction is heavily reverse biased by the larger battery, V_{CB}, making the collector very negative with respect to the base. The base–collector depletion layer is widened and this prevents holes crossing the junction from the collector to the base. But it in no way stops the flow of holes in the other direction. Thus, the holes injected by the emitter into the narrow base region rush across it, attracted by the heavily negatively charged collector, and enter the collector region.

However, not all of the holes leaving the emitter successfully travel the short distance across the base region to reach the collector and form collector current. A number of the holes, about one per cent or less of the total, collide with surplus electrons in the base region. The recombination of the holes and electrons in the base region leave the base region short of electrons; these are replaced from the negative plate of battery V_{BE}. If the base region electrons were not replaced, the base would become positively charged. The replacement electrons travel from the battery to the base region via the base connector. This constitutes a flow of conventional current from the base connector to the battery.

Clearly, then, the current emitted as holes from the emitter enters the base region and divides into two streams. The main stream passes straight through to the collector while a minor stream is lost down the base connector. The relationship between the three currents has been expressed earlier as Equation 8.1.

The ratio of the base current and the collector current is an important indicator as to how the transistor is likely to perform electrically. This ratio is called the *static current gain*, h_{FE}, where:

$$h_{FE} = \frac{I_C}{I_B} \qquad\qquad [8.2]$$

In words, Equation 8.2 says that if you measure the base current flowing, the collector current will be h_{FE} times larger. We have already seen above how, for every one hundred holes entering the base, say ninety-nine reach the collector and one is neutralized by an electron, so forming base current. Thus, we can expect h_{FE} to be 99/1, that is, about 99 times. In fact, the value quoted by transistor manufacturers varies typically in the range 100 to 300.

Several textbooks say that it is the base current, I_B, that controls the flow of the collector current, I_C. This is not strictly true. It is V_{BE} that controls I_C and it is h_{FE} that then decides the value of I_B.

The transistor operation described above is for the p–n–p transistor but it applies equally well to the n–p–n transistor provided that we first reverse the polarities of V_{BE} and V_{CB}, reverse the directions of the current arrows for I_E, I_B and I_C and wherever *holes* or *electrons* are mentioned, read the reverse: *electrons* and *holes*.

Figure 8.3 shows the circuit symbols for the p–n–p and n–p–n transistors.

8.3 The relationship between V_{BE} and I_C

We have seen that the forward biasing of the base–emitter junction is vital to the operation of the transistor. Whether the transistor be an

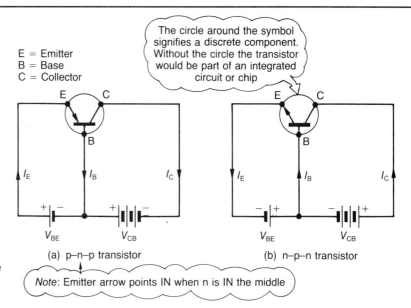

E = Emitter
B = Base
C = Collector

The circle around the symbol signifies a discrete component. Without the circle the transistor would be part of an integrated circuit or chip

(a) p–n–p transistor

(b) n–p–n transistor

Note: Emitter arrow points IN when n is IN the middle

Figure 8.3 *Bipolar transistor symbols connected for conduction to d.c. bias batteries*

n–p–n or a p–n–p, *in order to make the transistor conduct, the base voltage must be moved 0.6 V away from the emitter voltage towards the collector voltage.* This point is reiterated in Figure 8.4.

Figure 8.5 is a practical curve plotted from manufacturer's data. It shows how, for a p–n–p transistor, type BC 107 to 109, the collector current is controlled by the base–emitter voltage. Notice how there is virtually no conduction for values of V_{BE} from 0 V up to 0.6 V. But notice also how increments in V_{BE} from 0.6 V to 0.7 V are accompanied by dramatic current increases. In fact, for values of V_{BE} over 0.75 V, the collector current is tending to rise uncontrollably. The upper limit of collector current, say in excess of 300 mA, is where the transistor starts to overheat and ultimately burn out. It would have exceeded its safe power handling capacity.

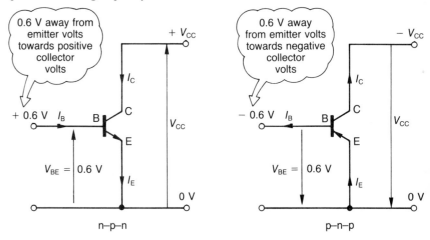

Figure 8.4 *For transistor conduction, the base voltage must be moved 0.6 V away from the emitter voltage towards the collector voltage*

V_{CC} = d.c. volts applied to collector

V_{BE} = d.c. volts between emitter and base

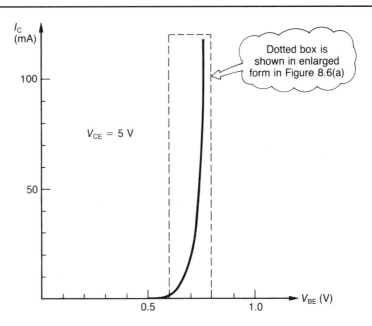

Figure 8.5 *Relationship between transistor output current, I_C, and input voltage, V_{BE}*

Transistors BC 107 to 109

Figures 8.6 and 8.7 are successive magnified views of selected sections of the main curve shown in Figure 8.5. A major significant feature of any one of these curves is its slope. This is determined by the ratio of change in the current flowing to the accompanying change in base–emitter volts. This ratio is given a special name, the transistor *mutual conductance*, g_m.

$$\text{Mutual conductance, } g_m = \frac{\Delta I_c}{\Delta V_{BE}} \text{ siemens} \qquad [8.3]$$

In Chapter 9, we shall see a use for g_m. It will be used for predicting the voltage gain of electronic amplifier circuits.

Example 8.1   *Figure 8.6(b) shows the I_c versus V_{BE} curve for a BC 108 transistor having a steady potential difference of 5 V between its emitter and collector connections (i.e. $V_{CE} = 5\,V$). Assuming the emitter to be held at earth potential and a collector current of 4 mA to be flowing, use the curve to determine (a) the d.c. voltage that is being applied to the base terminal and (b) the transistor's mutual conductance under these conditions.*

(a)   The value of V_{BE} at $I_C = 4\,$mA can be read off from point P from the curve at Figure 8.6(b) as 630 mV.

(b)   From Equation 8.3 we need to take the slope of the tangent at point P to find ΔI_C and the corresponding ΔV_{BE}. From the graphical work

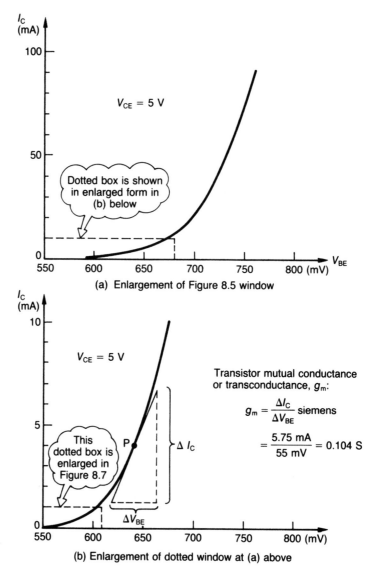

Figure 8.6 *Enlarged sections of transistor BC 107–109 characteristic curve of output current,* I_C, *against input voltage,* V_{BE}

shown on the curve in Figure 8.6(b), these work out as 5.75 mA and 55 mV respectively giving a value for g_m of 0.104 S or 104 mS.

Figure 8.7 is the expanded I_C versus V_{BE} curve for the BC 108 transistor in the region where conduction is just starting. This is where V_{BE} lies in the range 510 mV to 600 mV. In this limited range only, there is an approximate formula for estimating the value of g_m. This removes the need to refer to any characteristic curves or graphs and to this extent can be useful.

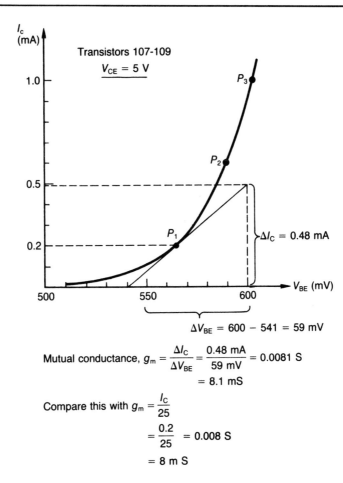

Figure 8.7 I_C versus I_{BE} near start of conduction

Inside figure:

I_c (mA)

Transistors 107-109
V_{CE} = 5 V

1.0

P_3

0.5

P_2

P_1

0.2

ΔI_C = 0.48 mA

500   550   600 V_{BE} (mV)

$\Delta V_{BE} = 600 - 541 = 59$ mV

Mutual conductance, $g_m = \dfrac{\Delta I_C}{\Delta V_{BE}} = \dfrac{0.48\ \text{mA}}{59\ \text{mV}} = 0.0081$ S

$= 8.1$ mS

Compare this with $g_m = \dfrac{I_C}{25}$

$= \dfrac{0.2}{25} = 0.008$ S

$= 8$ m S

The formula is:

$$g_m = \frac{I_C}{25} \text{ siemens} \qquad [8.4]$$

provided the value of I_C substituted into the formula *is the d.c. value in milliamps.*

Note that Equation 8.4 applies only to BJTs and not to FETs, which are discussed in Section 8.6.

Example 8.2 *A BC 108 transistor has a V_{CE} of 5 V and a steady collector current of 0.2 mA. Use Figure 8.7 to verify the approximate formula given by Equation 8.4.*

The point of interest on Figure 8.7 is P1. The tangent drawn to the curve at P1 has a slope of 8.1 mS.

Equation 8.4 gives a value for g_m of $0.2/25 = 0.008\,\text{S}$. Changing this into mS gives $g_m = 8.0\,\text{mS}$.

8.4 The transistor used as a switch

The transistor characteristic in Figure 8.5 shows that with $0\,\text{V}$ applied to its base, a transistor does not conduct; it performs as an open circuit. However, with only $0.75\,\text{V}$ on the base, the transistor conducts almost uncontrollably heavily; it is like a closed switch. Figure 8.8 shows a simple circuit which uses a transistor to switch a light on and off. The light requires a $50\,\text{mA}$ current to light it properly yet this level of current is controlled manually by SW1 making or breaking a current flow of only $0.5\,\text{mA}$ into the transistor base connector. This is made possible by transistor T1 having a current gain, h_{FE}, of 100. The pre-set battery produces the necessary V_{BE} to switch T1 into full conduction. The advantage of using the transistor as a solid state switch is the saving in switch contact wear which increases with the sparking and burning associated with making and breaking heavy currents. The motor-car ignition contact points of old have now been replaced by this type of solid state switching.

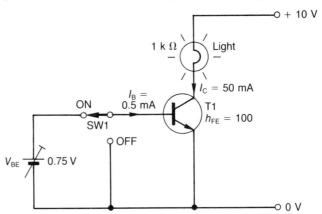

Figure 8.8 *n–p–n transistor used as a solid state switch*

8.5 The transistor used as a controller of current flow

Figure 8.9 shows the circuit arrangement for controlling the brightness of a light rather than simply switching it on and off. The controlling action is achieved by operating the transistor on the lower curved part of its characteristic – see Figure 8.5. The manual control is now exercised through a rotary potentiometer which taps off from the battery, V, a variable V_{BE} which is applied to the base of T1. With the potentiometer wiper in the $0.6\,\text{V}$ position, the light is about to glow and, as the wiper is rotated further clockwise, T1 is made to conduct more and the light to shine more brightly. T1 is simply a resistance, the value of which can be controlled by adjustment of V_{BE}. The greater V_{BE} is made, the less the effective resistance of T1, the greater the flow of I_C and the brighter the light.

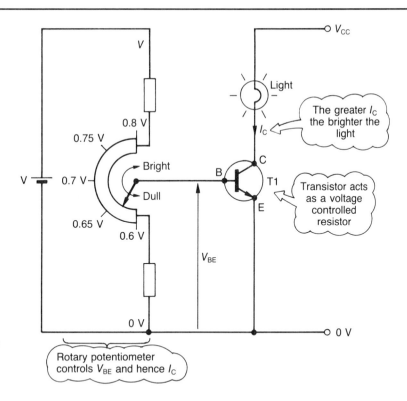

The greater I_C the brighter the light

Transistor acts as a voltage controlled resistor

Figure 8.9 *Transistor used as a current controller in a light dimmer circuit*

Rotary potentiometer controls V_{BE} and hence I_C

8.6 The junction field effect transistor (jFET)

One of the problems associated with the BJT is the large current that flows in the base connection in order to make the device work properly. We have seen that a transistor (this word is usually taken as meaning a BJT) having a current gain of $h_{FE} = 100$ and a collector current of 50 mA requires a corresponding base current of 0.5 mA. We shall see in Chapter 9 that the base terminal of a BJT can be the input to a voltage amplifier. This means that the voltage amplifier can need to draw 0.5 mA from its preceding circuit. On some occasions, drawing this amount of current from a circuit can change its operation or even stop it working altogether. The answer may be to use a different type of transistor for the amplifier input – one called a jFET or FET for short. This has the important property of requiring virtually no current at its input; it is purely voltage operated through what is known as an electric *field effect*.

Figure 8.10 shows a series of diagrams which help to explain the construction and operation of the FET.

Figure 8.10(a) simply shows the first stage of making an FET. It is a piece of n-type silicon connected across a battery. The n-type silicon conducts current by means of electron carriers and forms what is called an n-channel conductor. The electrons travel through the n-channel from the *source*, S, towards the battery positive terminal, V_{DS}, which is connected to the n-channel *drain*, D. Conventional current, I_D, flows from

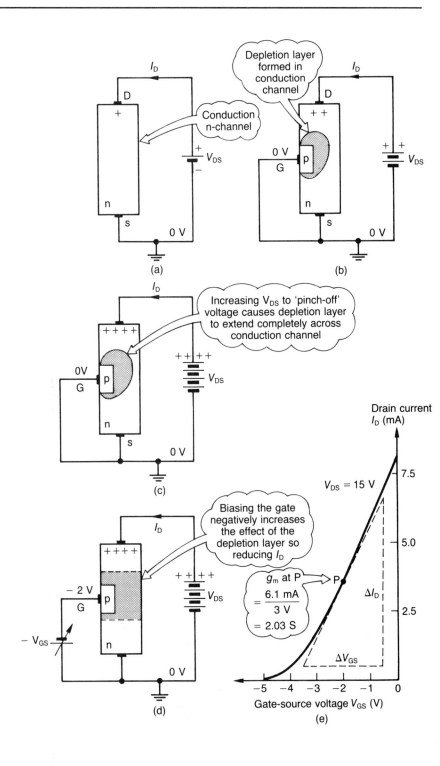

Figure 8.10 *Junction FET*

the n-channel drain terminal to its source terminal. The greater we make V_{DS}, the greater will be I_D.

Figure 8.10(b) shows the addition of a pellet of p-type silicon set into the side of the n-channel. The p-type pellet is called the *gate*. It is connected to the negative terminal of V_{DS} which is earthed to 0 V. With the gate at 0 V, and its surrounding n-channel material being attached to $+V_{DS}$ at some positive potential, a reverse biased p–n junction has been established. This is accompanied by the usual high resistance depletion layer which extends across and partly closes the n-channel carrying I_D.

Figure 8.10(c) shows the effect of increasing the battery voltage, V_{DS}. Initially it simply causes an increased flow of I_D, but since it also increases the reverse bias voltage across the p–n junction then it also increases the intrusion of the depletion layer into the n-channel. Eventually, the depletion layer extends right across the n-channel and further increase of V_{DS} causes no further increase in I_D. At this point, V_{DS} is said to be at the *pinch-off voltage*.

Figure 8.10(d) shows how the gate terminal is now used to control the flow of I_D. The p-type pellet is given a negative potential which further intensifies the depletion layer blocking the conducting n-channel. If, with V_{DS} about 15 V, the negative potential on the gate is made about -5 V, the n-channel becomes completely non-conducting and I_D stops flowing. Beyond pinch-off, the control of I_D has thus been passed to V_{GS} which is normally used to bias the gate to allow I_D to flow at half its maximum rate as set by the pinch-off voltage. The gate is thus said to operate in a *depletion* (or *reducing*) mode. It is of interest to note another difference between the BJT, which needs a kick of 0.6 V forward bias to start it conducting, and the FET, which requires a reverse bias voltage on its gate to reduce its inherent free flow of current.

Figure 8.10(e) gives a graphical representation of the relationship between the gate control voltage, V_{GS}, and the drain current, I_D. Note how the term mutual conductance arises as with the BJT. For the FET:

$$g_m = \frac{\Delta I_D}{\Delta V_{GS}} \text{ siemens} \qquad [8.5]$$

A point to note regarding the estimation of g_m for an FET: the approximate formula shown in Equation 8.4 applies only to BJTs and *not* FETs.

Compared with the BJT, the FET has a much lower g_m. The graphical example given in Figure 8.10(a) results in a g_m of only 2.03 mS. A BJT can produce ten times this figure, and more. We shall see in Chapter 9 how the low g_m of the FET gives any amplifier which uses it a lower gain than it would have using a BJT. However, the great advantage of the FET over the BJT is its low power consumption – it draws no current.

Figure 8.11 shows the circuit symbol for an n-channel jFET, or junction FET, of the type discussed above. jFETs having the conduction channel made from p-type silicon are available. The symbol for the n-channel type has the arrow on the gate reversed.

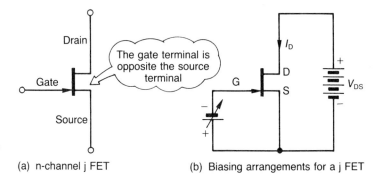

Figure 8.11 *jFET – symbols and biasing*

(a) n-channel j FET   (b) Biasing arrangements for a j FET

Exercises

8.1   A transistor has a collector current of $50\,\text{mA}$ and a current gain, $h_{\text{FE}} = 75$. Calculate the base and emitter currents.

8.2   Use Figure 8.6(a) to estimate the collector current flowing and the mutual conductance of the transistor when the base–emitter voltage is (a) $650\,\text{mV}$ and (b) $700\,\text{mV}$.

8.3   (a)   Use Figure 8.7 to determine graphically the value of g_{m} at points P2 and P3.

   (b)   Compare the values of g_{m} you have obtained in (a) above with those using the approximate formula.

8.4   Tests were carried out on a jFET to measure I_{D} at different settings of V_{GS}, all at a constant V_{DS}. The readings obtained are tabulated as follows:

V_{GS} (V)	0	−0.5	−1.0	−2.0	−3.0	−4.0	−5.0
I_{D} (mA)	8.0	6.0	4.75	2.4	1.0	0.3	0.001

   Plot a curve of I_{D} versus V_{GS} and use this to estimate the mutual conductance of the jFET when $V_{\text{GS}} = -3.0\,\text{V}$ and when $V_{\text{GS}} = -1.0\,\text{V}$.

9 Amplifiers

9.1 Introduction

A voltage amplifier is a device that accepts a small voltage waveform variation at its input and produces an enlarged, though possibly inverted, true copy at its output.

Figure 9.1(a) shows the block diagram of a voltage amplifier. The number of times that the output voltage, V_{out}, is greater than the input voltage, V_{in}, is called the *amplification*, *A*, or *gain*, *G*, of the amplifier. Figure 9.1(b) shows time-related input and output voltage waveforms for a 2 V peak-to-peak sinusoidal input voltage. The amplifier is taken as one that produces an inverted output voltage and has a gain of 2 times. The output voltage amplitude is therefore 4 V peak-to-peak:

$$V_{out} = A \times V_{in} \tag{9.1}$$

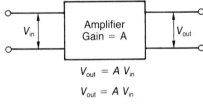

$$V_{out} = A\ V_{in}$$
$$V_{out} = A\ V_{in}$$

(a) Amplifier block diagram

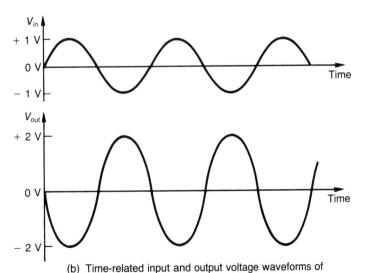

Figure 9.1 *Block diagram and voltage waveforms for a basic voltage amplifier*

(b) Time-related input and output voltage waveforms of a sinusoidal signal applied to an inverting amplifier having a gain or amplification factor of two.

Example 9.1 *A voltage amplifier has a gain of 120. Calculate the output voltage if the input voltage is 25 μV.*

The output voltage = amplifier gain × input voltage:

$V_{out} = 120 \times 25 \times 10^{-6} = 3 \text{ mV}$

From the above it is apparent that the voltage amplifier is a device which produces an output voltage which depends upon the size of the input voltage. In the next section we shall look at ways of building a circuit which will work in this way.

9.2 Using a simple resistor network for output voltage control

Figure 9.2(a) shows a 20 kΩ resistor, R, connected across a 12 V d.c. supply. The resistor passes a current of 0.6 mA and the full 12 V of the supply potential is dropped between points A and B.

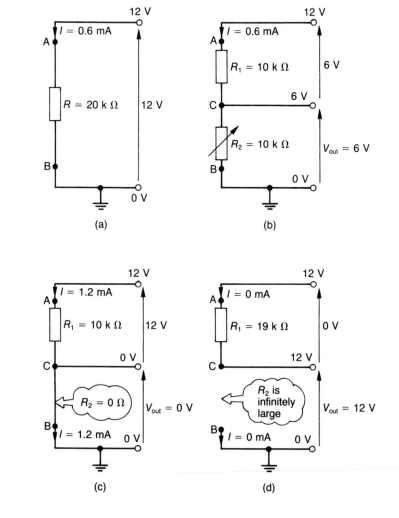

Figure 9.2 *A fixed resistor* (R_1) *in series with a variable resistor* (R_2) *can be used to produce a controlled variable voltage output,* V_{out}

If we now split the 20 kΩ resistor into two 10 kΩ resistors, there will be a 6 V drop across each 10 kΩ resistor giving a total of 12 V across the pair. If one of the 10 kΩ resistors is made variable we have the arrangement shown in Figure 9.2(b).

If the slider control of R_2 of Figure 9.2(b) is adjusted to make $R_2 = 0\,\Omega$, point C is effectively shorted to earth and so the output voltage, V_{out}, is also zero. This situation is shown in Figure 9.2(c). The full d.c. supply voltage of 12 V must now be dropped across the single 10 kΩ resistor, R_1. With one of the two 10 kΩ resistors having been effectively removed from the circuit, the current drawn from the supply has now doubled to 1.2 mA.

In Figure 9.2(d), the value of R_2 has been increased to an infinitely high value, effectively producing a break in the circuit where R_2 is fitted. Thus, current cannot flow through R_2 nor through R_1 because neither of them is in a completed loop circuit. Resistors R_1 and R_2 are said to be *open-circuited*. With no current flowing through R_1, there can be no volt drop across R_1 and the voltage at point C, the output voltage, V_{out}, must also be the full supply voltage of 12 V. Clearly, if we vary the resistance of R_2 between the two extremes of zero and infinity, the output voltage can be made to vary between zero and 12 volts. The next step is to make R_2 variable by voltage control.

9.3 Using a transistor as a variable resistor

Figure 9.3 is no more than Figure 9.2(d) redrawn but with the variable resistor, R_2, having been replaced by a transistor, T1. It will be recalled from Section 8.5 that the flow of current through a transistor can be accurately controlled by adjustment of V_{BE}. With the emitter terminal of T1 being earthed, the voltage applied to its base terminal is V_{BE} and any

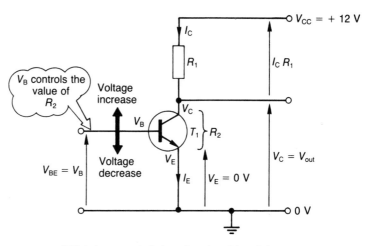

Figure 9.3 *A transistor can be used as a voltage controlled variable resistor to control* V_{out}

If V_B is increased, R_2 is reduced and $I_C \simeq I_E$ increases.
If V_B is reduced, R_2 is increased and $I_C \simeq I_E$ decreases.

variation of this voltage controls the effective resistance of T1 to current flowing through it. The larger we make V_{BE}, the more the current flow through T1 – see Figure 8.7. Now, the current that flows through T1 is approximately the same current that flows through resistor R_1, so the greater the current flow through T_1, the greater the volt drop across R_1. This is the same as saying that the greater we make V_{BE}, the greater will be the voltage drop across resistor R_1.

We can express the output voltage, V_{out}, mathematically as:

$$V_{out} = V_{CC} - \text{the volt drop across } R_1$$
$$V_{out} = V_{CC} - I_C \times R_1 \qquad\qquad [9.2]$$

Example 9.2 *If, in Figure 9.3, the collector resistor, R_1, 4.7 kΩ and the collector current, I_C, is 1.25 mA, calculate V_{out}.*

$$V_{out} = 12\,\text{V} - 1.25 \times 10^{-3} \times 4.7 \times 10^{3} = 6.125\,\text{V}$$

9.4 Building a simple voltage amplifier

We use the voltage, or *signal*, to be amplified as the voltage that controls the base–emitter voltage, V_{BE}, of a transistor. Then, the input signal will effectively control the current, I_C, flowing through the transistor which is also the same current flowing through the collector resistor, R_C. Figure 9.4 shows the circuit arrangement.

For example, suppose the current flow, I_C, through R_C is to be reduced by *decreasing* the voltage, V_{BE}, by a small amount. Then the consequent reduction in the volt drop across R_C results in a large *increase* in V_{out}.

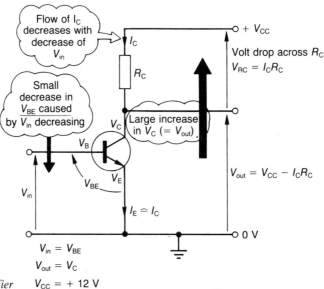

Figure 9.4 *Simple voltage amplifier*

Flow of I_C decreases with decrease of V_{in}

Small decrease in V_{BE} caused by V_{in} decreasing

Large increase in V_C (= V_{out})

I_C

R_C

V_C

V_B

V_{BE}

V_E

V_{in}

$I_E \simeq I_C$

○ + V_{CC}

Volt drop across R_C
$V_{RC} = I_C R_C$

$V_{out} = V_{CC} - I_C R_C$

○ 0 V

$V_{in} = V_{BE}$
$V_{out} = V_C$
$V_{CC} = + 12\,\text{V}$

Example 9.3 *Suppose a 60 mV fall in V_{BE} causes the current flowing through a transistor to reduce by 0.5 mA. If the collector resistor has a value of 4.7 kΩ, estimate the rise in output voltage.*

The change in volt drop across a 4.7 kΩ resistor caused by a reduction of 0.5 mA in the current flowing through it is:

$$4.7 \times 10^3 \times 0.5 \times 10^{-3} = 2.35\,\text{V}$$

Thus, we can conclude that the 60 mV *fall* in input signal is the cause of a *rise* in output signal of 2.35 V. Also, since a rise in input voltage is accompanied by a fall in output voltage, the amplifier output is an inverted version of the input. We can further conclude that the gain, G, of the amplifier is the change of output voltage, (ΔV_C), divided by the change of input voltage, (ΔV_{BE}), which caused it. The symbol Δ_1 means *a change of*. In this case the gain is:

$$G = (\Delta V_C)/(\Delta V_{BE}) = (-2.35 \times 10^3)/60 = -39.16 \text{ times.}$$

The minus sign does not in any way mean a reduction; it signifies that this amount of amplification is accompanied by *phase inversion*.

A further interesting point in Example 9.3 is the statement '. . . *the current flowing through a transistor to reduce by 0.5 mA.'* For this to be possible, the transistor must have had some current flowing through it before the application of an input signal to the base terminal. This leads into the next important point to understand with electronic amplifiers, that is, the need for d.c. biasing.

9.5 The need for d.c. biasing of amplifiers

In order to be able to increase or decrease the amount of current flowing through a transistor, the current flow should first be set at the mid-point between its maximum and minimum flow rates. If this is done, a symmetrical variation above and below this average value will be possible. In the case of our simple amplifier in Figure 9.4, *and in the absence of an input signal*, the base terminal voltage, $V_{BE} = V_B$, needs to be raised to, and held at, the necessary steady d.c. voltage required to cause half full-bore transistor current, I_C, to flow. With this value of current flowing, the value of collector resistor, R_C, which carries this same current flow, needs to be chosen so that the volt drop across it is just half that of the supply d.c. voltage, V_{CC}. This leaves the other half of V_{CC} to appear between the collector, the output voltage terminal and earth ($V_{out} = V_{CC} - I_C R_C$). The output voltage, $V_{out} = V_C$, in the absence of an input signal, is thus positioned halfway between its possible maximum, V_{CC}, and its possible minimum, earth (0 V). Clearly, equal excursions of the output voltage above and below this no-signal value is now possible. The values of V_{BE}, I_C and V_C in these no-signal mid-positions are known as *quiescent values* and are written as V_{BEQ}, I_{CQ} and V_{CQ} respectively.

Figure 9.5 shows the time-related input and output waveforms for a sinusoidal input voltage. The graphs represent typical values for an inverting amplifier having a voltage gain of 160 times. V_{BEQ} is held at 0.65 V and this causes the transistor to conduct sufficient current, I_{CQ}, to produce a 6 V drop across R_C leaving the collector voltage, V_{CQ}, also at

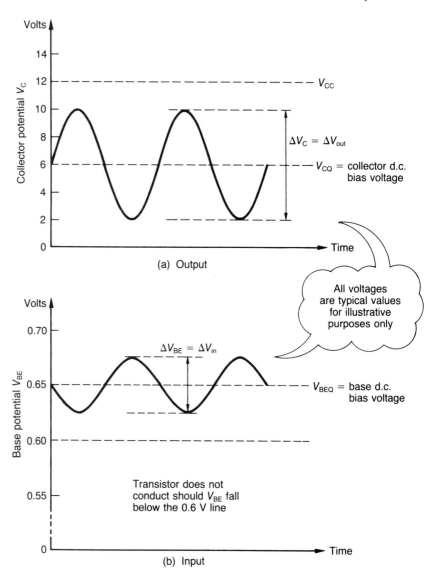

(a) Output

All voltages are typical values for illustrative purposes only

(b) Input

(c) Amplifier voltage gain, $G, = \dfrac{\Delta V_{out}}{\Delta V_{in}} = \dfrac{(10 - 2)\,V}{(0.625 - 0.675)\,V}$

$$= \dfrac{8\,V}{-\,0.05\,V} = -\,160\ \text{times}$$

Figure 9.5 *Time-related input and output voltage sinusoidal waveforms for the simple voltage amplifier shown in Figure 9.4*

6 V above earth. The input a.c. signal voltage, $\Delta V_{in} = 0.05$ V peak to peak, is superimposed upon the steady value of $V_{BEQ} = 0.65$ V making the base terminal swing sinusoidally between the limits of 0.0625 V and 0.0675 V. This causes an anti-phase sinusoidal swing of the collector voltage around $V_{CQ} = 6$ V of 8 V peak to peak, between the limits of 10 V and 2 V. The 8 V swing is, in fact, ΔV_{out}. Figure 9.5(b) illustrates the point that the amplitude of the input signal must not be so large as to swing the transistor into conduction cut-off. This would be the case if the base potential were taken below the 0.6 V level by the negative half-cycles of the input voltage. The result would be a distorted sinusoidal output waveform having flat tops by being limited at the $+V_{CC}$ level of 12 V.

The setting-up of the various quiescent d.c. conditions is called *biasing* the amplifier. We shall examine the biasing arrangements for two popular amplifiers: the common emitter amplifer and the common source amplifier.

9.6 Biasing the common emitter amplifier

Figure 9.6(a) shows the complete circuit of the common emitter amplifier. The amplifier is typically used as an audio amplifier and, as such, needs to produce an output which is an enlarged but otherwise a full, true copy of the input waveform. This is known as *Class A* working. The

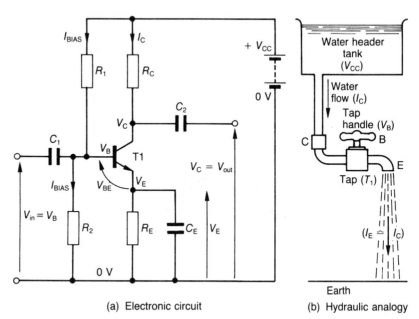

(a) Electronic circuit   (b) Hydraulic analogy

R_1, R_2 and R_E are bias resistors.

Figure 9.6 (a) The common emitter amplifier; (b) its hydraulic analogy

C_1 and C_2 block d.c. bias but pass a.c. signal currents.

C_E decouples R_E at a.c. signal frequencies.

amplifier has the name *common emitter* simply because the emitter terminal is common to both the input circuit loop and the output circuit loop. The biasing components are R_1, R_2, R_E and C_E.

R_1 and R_2 act together as a potential divider across the $+V_{CC}$ rail and earth. As a rule of thumb, assuming the current gain, h_{FE}, to be at least 100, the values of R_1 and R_2 should be chosen so as to have them pass a bias current, I_{BIAS}, that is at least one-tenth of I_C. This ensures that the small current drawn by the transistor base terminal through R_2 can be ignored compared with I_{BIAS} also flowing through R_2. The value of the resistor R_2 is selected so that I_{BIAS} flowing through it causes the top end of R_2 to be at the required voltage level, V_{BEQ}. This turns on $T1$ by the correct amount to cause I_{CQ} to flow and set the collector of $T1$ at V_{CQ}.

This action is analogous to the water tap in Figure 9.6(b) being turned on to its half full-bore water flow position. The water tap can now be opened further to allow a full flow of water from the header tank or closed off to stop the water flow completely. Precisely the same situation exists with $T1$ set with V_{BEQ} on its base. The transistor, $T1$, can have an increase in V_{BEQ} by a signal input, which will cause an increased flow of electrons through it. Equally, V_{BEQ} can be reduced to below the $0.6\,V$ level and $T1$ will cease to conduct.

R_E and C_E together form an automatic temperature stabilization circuit. An increase in temperature causes $T1$ to conduct more heavily which makes it warmer and a self-destruct spiral can result. To prevent this, R_E is fitted in the emitter lead such that with I_{CQ} flowing through it, the top of R_E is sitting at $V_{EQ} = +1\,V$. This means that V_{BE} must be set in excess of $1\,V + 0.6\,V = 1.6\,V$ in order that I_{CQ} flows. However, if I_{CQ} has an unwanted, gradual increase caused by a temperature increase, the voltage drop across R_E will increase correspondingly to more than $+1\,V$. Since V_B is held fixed by the junction voltage of R_1 and R, the increase in V_E effectively *reduces* V_{BE} and the transistor current I_C experiences a compensating reduction. The purpose of C_E is to decouple (effectively short-circuit) R_E for wanted, rapid changes in I_C caused by the input signal. C_E effectively earths the emitter at signal frequencies, so letting the amplifier amplify the signal which R_E, on its own, would inhibit.

C_1 and C_2 are in the signal input and output lines respectively simply to prevent the amplifier's d.c. biasing voltages being passed to any previous or succeeding circuits. The two capacitors have chosen values such that they will pass the a.c. signal frequencies but will block d.c.

9.7 Biasing the common source amplifier

Figure 9.7 shows the circuit arrangement for this type of amplifier, the voltage gain of which is much less than for the common emitter type discussed in the previous section. However, the common source amplifier has the great advantage of drawing virtually no input current. Figure 9.8 shows typical input and output voltage waveforms. Figure 9.8(a) shows the input a.c. signal, $V_{GS} = \Delta V_{in}$, as being a voltage variation around a quiescent gate bias voltage of $0\,V$. This is effectively a $-2\,V$ bias on the

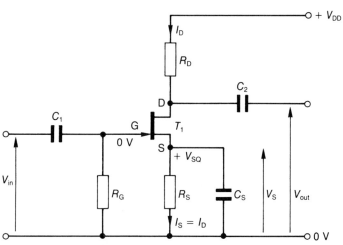

Figure 9.7 *Common source amplifier*

R_G and R_S are d.c. biasing resistors.
C_S decouples R_S at a.c. signal frequencies.
C_1 and C_2 block d.c. but pass a.c.

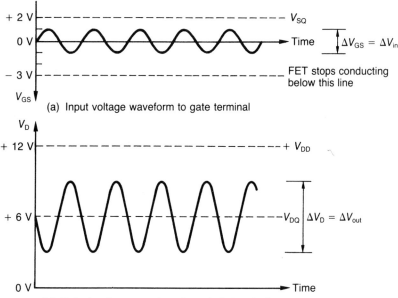

(a) Input voltage waveform to gate terminal

(b) Output voltage waveform from drain terminal

Figure 9.8 *Time-related sinusoidal input and output voltage waveforms for a typical FET common source amplifier*

(c) Voltage gain $= \dfrac{\Delta V_{out}}{\Delta V_{in}} = -\dfrac{6\text{ V}}{2\text{ V}} = -3$ times

gate terminal because the source terminal's quiescent voltage is arranged to be held up at $+2$ V by virtue of the steady quiescent current flow down through R_S. The effective negative bias voltage on the gate terminal is arranged to be correct for a quiescent current flow to give a quiescent drain output voltage of half V_{DD} for Class A amplifier working.

R_G is a very high value resistor (1 MΩ or more) which passes virtually zero current. Its purpose is to hold the gate terminal at 0 V in the absence of an input signal yet to allow the gate to follow the input signal variation.

C_S decouples R_S for signal frequencies leaving R_S to produce only the steady positive d.c. bias voltage, V_{SQ}. C_1 and C_2 are the usual 'signal voltage pass but bias voltage block' capacitors.

The voltage levels typified by Figure 9.8 show a voltage gain of only -3 times which compares very badly with the -160 times of the common emitter amplifier.

9.8 Testing the amplifier gain versus frequency response

For any voltage amplifier, the voltage gain is calculated by dividing the output signal voltage by the input signal voltage. In the case of the common emitter amplifier, Figure 9.6, the input signal voltage is ΔV_{BE} (R_E is shorted out by C_E at signal frequencies) and the signal output voltage is the signal voltage variation across R_C. This latter is $\Delta V_{out} = \Delta I_C R_C$.

Thus:

$$\text{Voltage gain} = \frac{\Delta V_{out}}{\Delta V_{in}} = \frac{\Delta I_C R_C}{\Delta V_{BE}} = \frac{\Delta I_C}{\Delta V_{BE}} R_C$$
$$= g_m R_C$$

If we put in the minus sign to signify the phase inversion in the single transistor amplifier, we can write:

Common emitter amplifier voltage gain $= -g_m R_C$ [9.3]

It can be shown that there is a similar relationship between the voltage gain of a common source amplifier, using a single FET, and the drain terminal resistor, namely:

Common source amplifier voltage gain $= -g_m R_D$ [9.4]

Example 9.4 *The amplifier circuit shown in Figure 9.6 has a value of 4.7 kΩ for R_C and transistor T1 has a mutual conductance of 40 mS. Calculate the output voltage for an input voltage of 20 mV.*

The amplifier voltage gain, A_V, is:

$A_V = -g_m R_C = 4.7 \times 10^3 \times 40 \times 10^{-3} = 188$ times

Therefore, the phase inverted output voltage is:

$V_{in} \times A_V = 188 \times 20 \times 10^{-3} = 3.73$ V

It might be supposed that this particular value of output voltage could be maintained for an input voltage of constant amplitude at any frequency. This could be investigated by producing what is known as the amplifier *frequency response curve*. To determine this response curve practically, a constant amplitude voltage, over a range of frequencies, is fed into the amplifier input and the resulting output voltage measured at each frequency step.

Figure 9.9 shows the block diagram of a suitable test equipment arrangement. Figure 9.10 shows a typical response curve. The ideal response is a horizontal line (shown dotted) representing a constant output voltage over the whole frequency range. On the other hand, the practical amplifier response is the curved solid line which approximates to the ideal over its centre part but falls, or *rolls-off*, at low and high frequencies. The reduced output at low frequencies is caused by the input capacitor; C_1 in Figure 9.7, for example. At too low a frequency, say below 30 Hz, the reactive impedance of the input capacitor is high and largely blocks the input signal. If there is a reduced input signal the output will also be reduced.

At the high frequency end of the response curve, the roll-off is due to both the input and output signals being leaked away to earth by the myriad of unwanted stray capacitances which exist between any pair of wires and any wire and earth. At high frequencies, say above 100 kHz, the total stray capacitive impedance is very low. These stray capacitances then allow most of the amplifier high frequency output to leak through them to earth; a reduced voltage at the amplifier output terminals proper is the consequence.

The range of frequencies over which the amplifier output voltage is at least 70.7% of its maximum is regarded as its working *bandwidth*. Because power is proportional to the square of voltage, the 70.7%

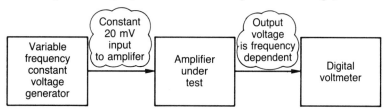

Figure 9.9 *Testing the frequency response of an amplifier*

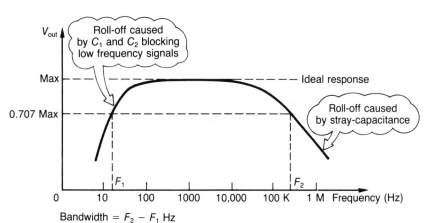

Figure 9.10 *Amplifier response curve*

Bandwidth = $F_2 - F_1$ Hz

F_2 and F_1 are respectively called the lower and the upper half-power points.

voltage points are also the points where the power output has fallen to half the maximum. On Figure 9.10, the bandwidth is the difference between f_1 and f_2 and these are also known as the *half-power frequencies*.

Exercises

9.1   Complete the following table for the block diagram of the amplifier shown in Figure 9.1:

	V_{in}	Gain, A	V_{out}
(a)	2 mV	200	?
(b)	?	300	2.6 V
(c)	4 µV	?	5.0 V

9.2   Suppose Figure 9.6 has component values as follows: $R_1 = 56\,k\Omega$; $R_2 = 12\,k\Omega$; $R_C = 4.7\,k\Omega$; $R_E = 1\,k\Omega$; g_m for $T1 = 40\,mS$ and $V_{CC} = +12\,V$:
   (a)   For component $T1$ there is an approximate formula for calculating g_m. Use this formula to estimate the g_m of $T1$ when the d.c. collector current is 0.0005 A.
   (b)   Explain what is meant by the terms *quiescent collector voltage* and *quiescent collector current*.
   (c)   If the quiescent value of I_C is 1 mA, calculate the quiescent values of V_C and V_E.
   (d)   Briefly explain the purpose and principle of operation of resistors R_1 and R_2.
   (e)   If V_{in} is 1 mV, calculate the value of V_{out}.
9.3   Figure 9.7 has component values as follows:
   $R_D = 4\,k\Omega$; $R_G = 1\,M\Omega$; $R_S = 2\,k\Omega$; $V_{DD} = 10\,V$:
   (a)   If the voltage between the gate and source terminals of $T1$ changes by 2.0 V and causes a change in the drain current of 3 mA, estimate the mutual conductance of T_1.
   (b)   What is the purpose of R_G?
   (c)   What is the purpose of R_S?
   (d)   If the quiescent value of I_D is 1 mA, estimate the quiescent voltages of V_D and V_S.
   (e)   If g_m is 1.5 mS and V_{in} is 3 mV, estimate the value of V_{out}.
9.4   A low frequency amplifier is tested to determine its output voltage versus frequency response curve. The amplifier input is fed with a constant 30 mV over a range of frequencies. The readings obtained are tabulated below:

f (Hz)	100	150	175	250	400	500	650	750	900	1100
V_{out}	1.0	3.0	4.0	5.7	6.0	6.0	6.0	5.4	3.7	1.0

Plot the response curve and use it as appropriate to determine:
- (a) the amplifier gain at frequencies of (i) 200 Hz, (ii) 600 Hz, (iii) 800 Hz
- (b) the lower and upper half-power frequencies
- (c) the amplifier bandwidth.

Appendix
Electronic circuit symbols

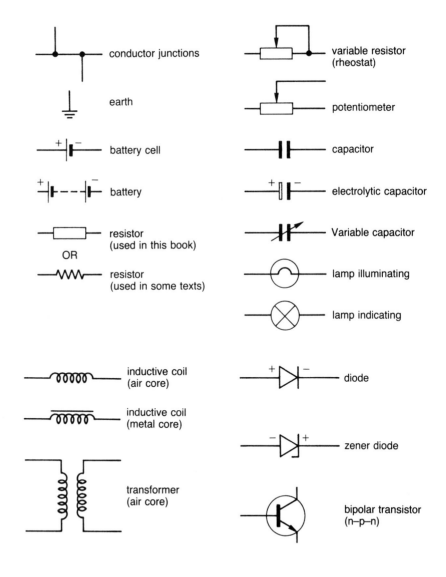

conductor junctions

earth

battery cell

battery

resistor
(used in this book)

OR

resistor
(used in some texts)

inductive coil
(air core)

inductive coil
(metal core)

transformer
(air core)

variable resistor
(rheostat)

potentiometer

capacitor

electrolytic capacitor

Variable capacitor

lamp illuminating

lamp indicating

diode

zener diode

bipolar transistor
(n–p–n)

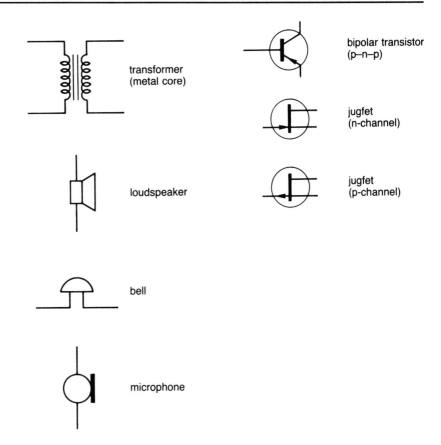

transformer
(metal core)

loudspeaker

bell

microphone

bipolar transistor
(p–n–p)

jugfet
(n-channel)

jugfet
(p-channel)

Answers to exercises

Chapter 2
2.1 4.8 h
2.2 0.38 Ah
2.3 7.2 V
2.4 10.18 V
2.5 91.9 mV
2.6 42.4 mV

Chapter 3
3.1 (a) 10 Ω
     (b) 1 A
     (c) 10 W
3.2 (a) 35 V
     (b) 25 V
     (c) 12.5 W
3.3 (a) 5.46 Ω
     (b) 1.1 A
     (c) 1.0 W
3.4 (d) 2 A
     (b) 76 V and 64 V
     (c) 1.2 A
     (d) 80 W
3.5 (a) L1 out, L2 and L3 brighter
     (b) Brighter, increased
     (c) Brighter
3.6 1.48 A
3.7 (a) 1.78 A, 0.445 A
     (b) 13.4 V
     (c) 17.9 W
3.8 17.6 W
3.9 1.61 W
3.10 1.40 A
3.11 2.14 A

Chapter 4
4.1 (a) 0.796 A
     (b) 15.9 mA

	(c)	1.59 mA
4.2	(a)	62.8 A
	(b)	126 mA
	(c)	19.7 mA
4.3	(a)	47.66 Ω
	(b)	0.21 A
	(c)	1.19 W
4.4	(a)	87.81 Ω
	(b)	2.62 A
	(c)	34.9°
	(d)	131 V
	(e)	143.7 V
4.5	(a)	27.27 Ω
	(b)	1.83 A
	(c)	63.89°
	(d)	0.44
4.6	(a)	128.4 Ω
	(b)	3.5 A
	(c)	1575 VA
	(d)	833 W
4.7	(a)	82.19 Hz
	(b)	22 Ω
	(c)	1.13 A
4.8	(a)	33.33 Ω
	(b)	0.056 nF
4.9	(a)	3.45 A and 2.36 A
	(b)	1.97 A
	(c)	14.36°
	(d)	126.9 Ω
	(e)	0.968
4.10	(a)	3.09 A, 0.537
	(b)	1.66 A, 36 mF

Chapter 5

5.1	(a)	239.6 V
	(b)	5.99 A
	(c)	5.99 A
5.2	(a)	415 V
	(b)	10.38 A
	(c)	17.97 A
5.3	(a)	3200 W
	(b)	9600 W
5.4	(a)	274.8 V
	(b)	475.97 V
5.5	(a)	5.82 A
	(b)	10.08 A
	(c)	5081 W
5.6		4805 W, 0 707 lagging

5.7 3286.8 W, 34.1° leading
5.8 (a) 3.63 A
(b) 10.9 A

Chapter 6
6.1 (a) 16.22%
(b) 3.74%
6.2 (a) 2.04 Ω
(b) 900 Ω
6.3 (a) 0 V, 0 Hz
(b) 15 mV, 0 Hz
(c) −6 V, 0 Hz
(d) 15 V, 62.5 Hz
(e) 15 mV, 50 kHz
(f) 6 V, 8.33 Hz
(g) 15 V, 83.3 Hz
(h) 15 mV, 66.67 kHz
(i) 6 V, 12.5 Hz

Chapter 8
8.1 0.67 mA, 50.67 mA
8.2 (a) 6 mA, 146 mS
(b) 22 mA, 596 mS
8.3 (a) 23.3 mS, 40.6 mS
(b) 24 mS, 40 mS
8.4 1 mS, 2.73 mS

Chapter 9
9.1 (a) 0.4 V
(b) 8.67 mV
(c) 1250
9.2 (a) 20 mS
(c) 7.3 V, 1 V
(e) 188 mV
9.3 (a) 1.5 mS
(d) 6 V, 2 V
(e) 18 mV, phase inverted
9.4 (a) (i) 150 (ii) 200 (iii) 167
(b) $f_1 = 180$ Hz, $f_2 = 870$ Hz
(c) 690 Hz

Index

$y=x^2$

$x=1$

$y=x^2$

$y=x$

$y=3^2x$